Henry Neville Hutchinson

**Prehistoric Man and Beast**

Henry Neville Hutchinson

**Prehistoric Man and Beast**

ISBN/EAN: 9783743383999

Manufactured in Europe, USA, Canada, Australia, Japa

Cover: Foto ©berggeist007 / pixelio.de

Manufactured and distributed by brebook publishing software (www.brebook.com)

Henry Neville Hutchinson

**Prehistoric Man and Beast**

# PREHISTORIC MAN
# AND BEAST

BY

REV. H. N. HUTCHINSON, B.A., F.G.S.

AUTHOR OF 'EXTINCT MONSTERS' ETC.

'Large are the treasures of oblivion. Much more is buried in silence than recorded.'—SIR THOMAS BROWNE: *Religio Medici*

WITH ILLUSTRATIONS BY CECIL ALDIN

LONDON
SMITH, ELDER, & CO., 15 WATERLOO PLACE
1896

[All rights reserved]

# PREFACE

By SIR HENRY H. HOWORTH, K.C.I.E., M.P., F. .S., ETC.

THERE are few people so dull or so frivolous that they are not interested in the beginnings of human history. We all long at times for some Aladdin's lamp which shall enable us to thread our way through the fogs and mists with which time has shrouded the old story. The traditions of men have responded to this yearning in various ways, some allegorical and some prosaic; while in some, again, the too fantastic story has been worked over and made to look more reasonable by later hands.

Meanwhile, many and various students have tried to reach the same goal by harder roads, and these have been very differently paved. Some have taken human language in all its various forms, and have analysed and compared these forms with the object of reaching its earliest stage, and thus trying to measure the capacity, the arts, and the progress of man when he was still in his cradle by analysing his speech. Hitherto this method has furnished us with rich harvests of interesting facts, and opened up delightful vistas for the future student to explore, but they have not carried us far along the particular road which we would travel. Go back as far as we like, we seem to find human language developed in its most elaborate forms, with those refinements and that rich vocabulary which we associate with the highest development of thought and knowledge. This is not all. These same languages seem as far apart from each other at the earliest stage to which we can carry them as they do now—Sanskrit and Chinese, Babylonian and Egyptian. These are samples only of what we mean.

While some have been trying to tap the fountain of language, others have done the same with the history of religion and mythology. Few things which engage the thought of human beings are so conservative as ritual and mythology. They deal with the most sacred thoughts that occupy men, where it is difficult to move without exciting popular prejudice, and where, therefore, violent changes are nearly impossible. It is not surprising, therefore, that in this field also many and brilliant advances have been made in recent years which have almost revolutionised the science of comparative mythology; but here also the progress towards our final goal has been slow and tedious, and we see the roadway still occupied by bramble and thorn, and realise what patient and hard work will have to be done for many a long day before the story is completely told.

It is the same with archæology, that is to say, the inquiry into the beginning and early progress of human arts, a subject which has occupied a whole army of workers in many countries for many years past. Here also a vast deal that was once obscure and confused now seems orderly and fairly plain; but great gaps and blank places mark how much still remains to be known, how much is imagination and hypothesis, and what promise there is that if we go on patiently working we shall eventually, if we do not quite complete a story which is so largely lost, go far towards that desirable end.

It is a fortunate thing when, in subjects like this, there are found at times those with patience and knowledge and literary skill, gifted also with imagination and the power of graphic description, who will undertake to gather and put together the scattered threads of inquiry into a continuous web and pattern, digging here and digging there in those dark holes and serpentinous caverns in which the specialist works, and make everyday readers acquainted with what has been done. This is the work which Mr. Hutchinson, a practised pioneer in this field, has essayed to do in the following pages. In them he has discussed many burning archæological questions and traversed many fields where men still disagree. He has told the story simply and intelligently. He has let people see where the ground is firm and where it is slippery, and he has skilfully constructed a chronicle of human affairs for times before history properly begins. He has begun with the discussion of the oldest facts we possess about man, and has continued the story down to historic times properly so-called, and has presented the English reader with a balance-sheet showing our present state of knowledge upon a delightful and engrossing subject of human inquiry, and he cannot fail to secure a large number of readers. My only hesitation in speaking of the book in suitable terms is the exaggerated language which he has been good enough to use about some of my own scattered and desultory gleanings in this field. We cannot agree about all matters; we each of us have our own personal heresies and prejudices; but I can frankly say that I know of no recent book in the English language on this subject so full of information, so graphic, and so inspiring as this one.

A few words should also be said about the illustrations. Men and boys who visit museums and see the broken and decayed remains of our ancestors seldom have the imagination or the knowledge to complete a mental picture of the living men and their times of which these are débris. It is necessary, therefore, for someone who combines imagination and deft fingers to represent for us the actual life as nearly as it can be reached. This has been done in the present case with a great deal of pains and care, and Mr. Hutchinson and his artist must be congratulated upon them.

HENRY H. HOWORTH.

30 COLLINGHAM PLACE, LONDON.

# AUTHOR'S PREFACE

THIS is a book intended for everybody, not for the specialist either in Geology or Archæology. Such may find here some accounts of recent researches, but the book is not written for them. Its purpose is to give in a simple style some of the most interesting results arrived at of late years by the two diligent armies of workers who have been labouring in those pleasant fields of research and speculation. The writer has endeavoured to place himself in the position of an interpreter, not of a Brahmin speaking a language unknown to the people.

The subject of his two former works, 'Extinct Monsters,' and 'Creatures of Other Days,' written with the object of tracing the course of development of animal life on the earth, has naturally led the writer on to the subject of Man himself—the final product of countless ages of Evolution, the 'beauty of the world, the paragon of animals,' and the highest manifestation of Creative Power that has yet appeared thereon.

As the late Poet Laureate wrote:

> First the Monster, then the Man;
> Tattooed or woaded, winter clad in skins,
> Raw from the prime and crushing down his mate;
> As yet we find in barbarous isles, and here
> Among the lowest.
> *The Princess.*

Another suggestion tending towards the same end has been afforded by those clever and humorous 'Prehistoric Peeps' which have for some time past given so much delight and amusement to readers of 'Punch:' the animals which figure therein with such quaint effect are, for the most part, taken, at the writer's friendly suggestion, from the two books above mentioned, and thus have served the double purpose of giving variety to the pictures and of being also partly responsible for the present undertaking. This, we venture to hope, may, as it can hardly be expected to amuse, yet perhaps afford some instruction to those who desire to eat of the tree of knowledge, which, as the old Greek philosopher said, is the same thing as Virtue. Yet, if some should doubt the truth of this saying, we can at least console ourselves with the thought that any one who reads our book will, at all events for the time being, be kept out of mischief! We therefore thank our friend the 'Special Prehistoric Artist' of that great and powerful person 'Mr. Punch' for his kind, albeit unintended, suggestion.

It is curious to observe how theories once put forward, if only they are striking and somewhat sensational, lay hold of the popular imagination, and remain fixed there long after they have been practically abandoned; a case in point will be found in Chapter IV., where it is shown that the theory, even now so popular and so attractively handled by lecturers, of the Great Ice-sheet, or Polar Ice-cap, is nothing less than a monstrous fiction, which, in spite of the clever special pleading of Prof. James Geikie and Sir Robert Ball, has done a great deal of harm. Let us hope that it will never appear again, for it seems to have had a most chilling and deadening effect on the minds of some workers and writers, whose brains, relieved of this 'nightmare,' may be expected in time to blossom forth and ripen into more fruitful, because more true, thoughts! And with this, another baneful growth is also withering up—we mean the much talked of Astronomical theory of the Ice Age.

A chapter is devoted to each of these subjects. In both cases the evidence breaks down all round; and if any of our readers object to see their favourite theories knocked on the head, it is only necessary to remind them that Truth, being stronger than Fiction, must prevail. Sir Henry Howorth, in his 'Glacial Nightmare and the Flood,' has exposed the fallacies on which they are founded.

The antiquity of man, another subject of the highest interest, is discussed in two special chapters in which certain considerations brought forward by distinguished men of science are explained, to show that the oldest men of which we have any sure knowledge at all, instead of belonging to a time of vast and highly remote antiquity, come within at least a more 'measurable distance.' Only those few geologists who still fondly cling to the Astronomical theory of the Ice Age can possibly bring themselves to believe that the men who hunted the mammoth and the reindeer in the older Stone Age lived at a time remote from the present by some 80,000 years or more! On this subject the writer has gladly followed the careful and moderate estimates of the late Sir Joseph Prestwich, to whom he is grateful for kind counsel.

But when we come to consider *civilised* man, the case is very different; here we find that the whole tendency of modern research is to throw back civilisation to periods much more distant than was dreamed of not very long ago. Civilised man appears to have lived in Egypt about 10,000 years ago. Even in ancient Britain there was a good deal of civilisation before the time of the Roman conquest. Julius Cæsar failed to do justice to our Britannic ancestors, and we must no longer regard them as naked savages; with the help of the artist, we have endeavoured to depict the ancient inhabitant of these isles clothed and, so to speak, in his right mind.

Perhaps one of the most interesting results of recent research is the discovery that the 'Fairies' were a real people! Various writers have come to the conclusion that a dwarf population akin to the Lapps, and hailing from Finland, were the actual people who lived in the 'Fairy knowes,' or underground megalithic structures, and became in due time the fairies of Folk-lore. A special chapter has been devoted to this subject.

Stonehenge is discussed in Chapter XII., with the object of showing that it can no longer be regarded as the work of the Druids. Other stone-circles in Britain, in Europe, and Asia are dealt with in the previous chapter, and the facts there stated are intended to pave the way for an impartial consideration of our greatest British monument. The writer is inclined to believe that it is *not* a temple, but of a purely memorial character, though at the same time sacred, because connected with the cultus of departed spirits. He has even ventured to suggest that it might have been erected by the dwarfs. Certain analogies with the chambered mounds and dolmens seem to lend some weight to the idea that possibly 'Fairies' were the real builders of Stonehenge!

The whole subject of stone-circles is still very obscure, and the above theory has been put forward chiefly with the object of producing further discussion. With regard to the astronomical theories, so popular with some writers, it has not been thought worth while to discuss them here, since there is little or no evidence to sustain those ingenious speculations.

Nowadays, discoveries follow one another so rapidly that a book of ten or fifteen years ago on Prehistoric Man becomes somewhat out of date. An endeavour has been made in the following pages to give to the general reader some idea of the latest discoveries and the most recent conclusions of geologists and archæologists. But the literature of the subject is now so enormous that no one person can master it all. A selected list of important works will be found at the end of the book.

There are no small difficulties in the way of making good 'restorations' of prehistoric men, especially when we go back to the Older Stone Age, and the writer is far from claiming absolute accuracy for those in this book. But if they succeed in conveying to the reader only a general idea of our ancestors' ways and doings, he will consider that the attempt was worth making. They are the first restorations that have appeared in any English work, and it is hoped that their defects will be judged leniently.

Great credit is due to the artist, Mr. Cecil Aldin (of the *Illustrated London News*), for the admirable manner in which he has carried out the writer's suggestions. His illustrations, instead of being mere scientific diagrams, are thoroughly artistic and vivid pictures, which would well bear reproduction on a larger scale. In order to avoid mistakes as far as possible, the writer has submitted the rough sketches and drawings to several archæologists and scientific men. Sir William Flower, Sir Henry Howorth, Dr. Henry Woodward, Mr. A. Smith Woodward, Canon Greenwell, Prof. M. Flinders Petrie, Mr. C. H. Read, and Prof. J. McKenny Hughes have made some valuable suggestions. Dr. Sophus Müller, of Copenhagen Museum, kindly sent his outline restorations of a man and a woman of the Danish Bronze Age from his work 'Voroltid,' which were very helpful to both author and artist in working out Plate VIII.

The writer has received much kind help both from geologists and archæologists, for which he is very grateful. Sir Henry H. Howorth has taken an interest in this work from the beginning, and most kindly read all the proofs, making at the same time valuable notes and comments, which the writer was not slow to use. He has also written a far too flattering Preface, which, coming from so learned an author and so kind a friend, deserves the writer's warmest thanks. The writer's thanks are also due to Mr. Arthur Smith Woodward, of the Natural History Museum, and to Mr. A. H. Scott White for reading proofs, and otherwise lending a helping hand.

LONDON, 1896.

*Note.*—Societies or institutions wishing to arrange for lectures by the author on the subject of this book, or of others which he has written, may apply to the Lecture Agency, Outer Temple, 225 Strand. Lantern slides (copyright) of all the illustrations may be obtained from Messrs. Newton & Co., 3 Fleet Street.

# CONTENTS

## INTRODUCTION

PAGE

Knowledge of the Ancients—Modern theories anticipated—Lucretius, Job, Babylonian cycles of time—Evolution, and the old figure of the Tree of Life—Darwin and Lamarck—Aristotle's 'Principle of Perfection'—Man's lowly origin—The account in Genesis implies evolution—Order of events—The records of Archæology—Modern primitive races of great service—Archæological periods . . .   1

## PART I

### THE MEN OF THE OLDER STONE AGE

### CHAPTER I

#### OUR EARLY ANCESTORS

Birthplace of Man unknown, but probably in the Old World—High civilisation seems connected with a temperate climate—Lucretius on human progress ('De Rerum Naturâ')—Discovery of M. Boucher de Perthes at Abbeville that man was contemporary with the extinct animals of the river-gravels—Anticipated by Dr. Schmerling in Belgium—Rev. J. MacEnery arrives at same conclusion—Types of flint implements from the gravels—How used—Distinction between Palæolithic and Neolithic implements—Their high antiquity—Geographical changes—Contorted drift—Plateau implements of Mr. Harrison—Prestwich's theory of their pre-glacial age—Some geologists do not accept them as implements—Human remains from Spy—Huxley's description—Accepted as Palæolithic—Human remains from Galley Hill, Northfleet—Mr. E. T. Newton's description—Their age not settled—Mr. Worthington G. Smith's discovery in Suffolk—Palæolithic floors—Flint-cores—Primæval man according to Mr. Worthington G. Smith—A fanciful picture—The now extinct Tasmanians appear to have been the nearest approach to Palæolithic man . . . . . . . . . . . . 15

## CHAPTER II

### ANCIENT CAVE-DWELLERS

Superstitions about caves—Discoveries of Schmerling and Cuvier discredited—Formation of limestone caves—Dean Buckland and his explorations—His diluvial theory—Kirkdale cavern, once a hyæna den—How he proved this—Wookey Hole—Animals that inhabited the neighbourhood—Description of Plate I.—Kent's cavern—The fauna of the caves—Three different groups: Southern, Northern, and Temperate—Were they contemporaneous?—Theory of annual migrations—J. Geikie's theory of mild inter-glacial periods—Evidence that the hyæna was contemporary with the reindeer—Climatic conditions of the time like those of Switzerland—Does boulder clay occur in the caves *in situ*?—Dr. Henry Hicks on Cae Gwyn cave—View of Prof. T. McKenny Hughes—The Victoria cave—Mr. Tiddeman's view—Difficulty of interpreting cave-deposits . . 40

### CHAPTER III

### THE REINDEER HUNTERS

Rock-shelters of the Périgord—First explored by Messrs. Christy and Lartet—Books on the subject—Remains of feasts—Localities explored—Immense number of horses' bones—Reindeer abundant—Artistic engravings of animals—A varied fauna—Climatic conditions—Weapons, utensils, and ornaments—Specimens thereof in British Museum—Use of paint—The people probably were hairy—Description of Plate III.—Tents or huts?—The art of making pottery unknown—*Bâtons de commandement*, theories of their use—M. de Mortillet's subdivisions of the Older Stone Age: Chellean, Mousterian, Solutrean, Magdalanean—Separation of Older and Newer Stone Ages, in several ways—Supposed palæolithic skulls—Supposed great gap—Prof. Boyd Dawkins' description of the reindeer hunters—The Cro-Magnon race, M. de Quatrefages' view probably mistaken—List of skulls which are probably *not* Palæolithic (as some have maintained)—Reported discovery at l'Ain—Another from the Loess of Moravia—Burials probably did *not* take place among the reindeer hunters—Survivals from the Stone Age—How a flint-knife is made—Uses of Palæolithic implements—Striking a light—How was cooking invented?—Customs of modern savages—Picture by Tacitus—Difficulty of ascertaining the religious ideas of savages—Remarks of Prof. Max Müller—Prof. Tylor on 'Animism' . . . . . 62

# CHAPTER IV

## THE MYTH OF THE GREAT ICE-SHEET

Erratic boulders—Scottish legends about them—Theory of the Polar Ice-cap—The Glacial Nightmare—Efforts of Sir Henry H. Howorth and others to upset the theory—Boulder clay, how formed—Evidence of pressure—Roches moutonnées—Abrading action of ice much exaggerated—Intercalated layers of sand and gravel—Marine shells in boulder clay—Yet this deposit was *not* generally formed under the sea, as formerly taught—Iceberg theory abandoned—'The Great Submergence'—Changes of opinion often necessary—Historical survey of the subject : Swedenborg, De Saussure, Hutton, Sir James Hall, Sedgwick, De la Beche, Murchison, Charpentier, Goethe— Agassiz got his theory from Schimper—British glaciers—Absence of far-travelled stones in boulder clay—Danish explorations—Scandinavian boulders on our East coast, how explained—Sir H. Howorth —North American evidence—Palæolithic man appears to have been here during the Ice Age—J. Geikie's 'inter-glacial periods' unproved —Great local floods—The Loess, theories of its formation—View of Prestwich—Great numbers of bones in some caves—Chaldean version of the Flood—Howorth on the Flood as the cause of extermination of the mammoth—The supposed 'Great Gap' . . . . 84

# CHAPTER V

## CHANGES IN CLIMATE AND THEIR CAUSES

Remarkable changes in flora of Greenland since Miocene times, when the climate was warm—The Grinnell Land lignite—Jurassic warm-water shells at Parry Islands—Speculations as to the cause of such change—Cooling of the earth—State of the atmosphere—Colder regions of space—Meteorites—Comets—Variable stars—Sun-spots— Possible greater size of the sun—Tyndall on production of glaciers— Earth's axis of rotation—Obliquity of the Ecliptic—Wandering of the poles, hardly possible—George Darwin's and Lord Kelvin's opinions—H.M.S. 'Challenger's' cruise and its results, bearing on the doctrine of the permanence of ocean basins—Rev. E. Hill, Dr. Croll, and Professor George Darwin on changes in earth's axis —Croll's astronomical theory explained—Variations in the eccentricity of the earth's orbit—Periods of high eccentricity—Herschel's theorem—Precession of the Equinoxes—Revolution of the apsides— Effect of winter in aphelion—And of summer in aphelion—Herschel

a

was doubtful of the supposed effects—Croll's arguments about the trade-winds—Objections to the theory—The astronomical data insufficient—No geological evidence worth considering of other glacial epochs, such as required by the theory—No evidence of alternations of warm and glacial periods in the two hemispheres—Recentness of glacial period – A more likely theory is that of simple elevation—Evidences of submergence –Glacial succession of strata . . 107

## CHAPTER VI

### THE ANTIQUITY OF MAN

Man *may* have originated in Miocene period—What is *known* different from what is only probable—Evidence of the antiquity of civilised man—Theory of Mr. Wm. Peck with regard to the date at which the signs of the Zodiac originated—Annual path of the sun among the stars—Each figure of the twelve signs of the Zodiac supposed to refer to the season, or natural events of the time at which the sun was in that group—Sphere of Eudoxus at Rome—Eudoxus travelled in Egypt—Changes in position of some constellations—Many of the signs seem to refer to the seasons in Egypt, and the occupations of the people at those seasons—Or to the weather, state of the Nile, &c.—Each sign taken in turn and explained—Geological data, from which *some* idea of the antiquity of the human race may be gained : (1) Delta of the Nile ; (2) Delta of the Mississippi ; (3) Niagara Falls ; (4) Excavation of river valleys ; (5) Pedestal-boulders ; (6) Flow of Greenland ice ; (7) Formation of stalactite in caves ; (8) Time required for the growth of peat-beds—Each of these questions briefly discussed—They all agree in proving that Croll's minimum of 80,000 years since the Glacial period cannot be accepted . . . . 128

## CHAPTER VII

### ANTIQUITY OF MAN (*continued*)

(4) High-level gravels of the Thames—Prestwich considered this river had excavated its valley to a depth of about eighty feet since Glacial times—Theory of uniformity has been stretched too far—Observations on the Mississippi to calculate the rate of denudation all over its valley—About one foot in 6,000 years—This is too slow a rate to apply to the Thames gravels—Prestwich's later opinion was that the high-level river-gravels were of Glacial age, and *not* post-glacial—(5) Pedestal-boulders—How had they got there?—Not left by icebergs

—Nor washed out of boulder clay—But left there by the old glaciers—As explained by Professor T. McKenny Hughes—These boulders are standing slightly above the level of the limestone around them—This slight elevation of 12-20 in. represents the amount of subsequent denudation—It seems impossible that in 80,000 years so little denudation was effected !—Therefore Croll's calculation will not stand—Wonderful preservation of glacial striæ—This gives a like argument for the recentness of Glacial times—Did man live in Tertiary times ?—Views of Dr. Keith Falconer—Supposed discovery of Dr. Noetling in Burmah—Mr. R. Oldham gives the true explanation—The flint-flakes were *not* Miocene, but had been washed down on to Miocene beds, and so *appeared* to belong to them—Other supposed discoveries of Tertiary man have shared the same fate—With possible exception of the Java skull—The case of the Thenay implements—They were Pleistocene, but had slipped down into a pit, and got mixed with Tertiary fossils—The case of implements found at St. Prest—The marks on the bones probably those of fishes' teeth—Professor T. McKenny Hughes on similar marks on a Plesiosaurus bone—Other cases—The recently found Java remains—Huxley's statement of the case thirty years ago—Dr. Dubois discovers human remains in volcanic tuff in Java in 1891—The remains consist of a calvarium, a femur, and one tooth—But not all found together—The skull seems too large for the femur—The skull very degraded and ape-like ; its capacity 1,000 cubic centimetres—The opinions of anatomists—They do not agree—Some say the remains are those of an ape—Some say they are human—Others that they represent a 'missing link'—Others, again, think they represent a diseased human being—The question of the geological age of the strata does not appear to be settled . . 141

# PART II

## *MEN OF THE LATER STONE AGE AND BRONZE AGE*

### CHAPTER VIII

#### DWELLERS ON THE WATER

Herodotus on the lake-dwellings of Pæonians—Hippocrates' description of the people of Phasis—Pliny on the Frisians—Lake-dwellings still used in various parts of the world—British 'crannoges'—Sites of Swiss lake-dwellings—Some of the Stone Age—Some Bronze Age—

A few Iron Age—Characteristics of Stone Age settlements—Researches of Dr. F. Keller—Literature of the subject—Made of piles and wattlework—Roofs of straw or reeds—Steinbergs—Varieties of construction—External shape—Discovery at Schussenried of a settlement preserved in peat—Dr. Munro's description—Stone-axes—Importance of jade implements, as probably indicating commerce with the East—The boring of stone-hammers—The Robenhausen settlement (Stone Age)—Hearths—Domestic Animals—Fish-hooks, bows, arrowheads, spearheads, saws, and awls—Clothing—Flax used for spinning—Spindle-whorls—Wickerwork—Personal ornaments—Pottery hand-made—Head-rests, hair-pins, toys, bridle-bits—Wild and domestic animals of Stone Age and Bronze Age—Corn and cakes—Fruits and vegetables—Hemp not found—Did a new race come in the Bronze Age?—Conflict with the Romans in Bronze Age—Writing unknown in the Bronze Age—Battlefield of Tiefenau—Objects of pure copper—Iron Age settlements of Marin and La Tène—Frank's 'Late Keltic' Period—The founders of the lake-dwellings, probably the Neolithic and pre-Keltic Race—Burials in the 'contracted position'—Tombs at Auvernier—Cremation *and* inhumation—Stone Age burials at Chamblandes—British Crannoges—They are much later—Examples at Lough Lane (Ireland), the Loch of Forfar, Lochindorb and Loch Cannor (or Kinord), Scotland—Canoes made of tree trunks. . . . . . . . . . 167

## CHAPTER IX

### ABODES OF THE LIVING AND THE DEAD

The kitchen-middens of Denmark—Hearth-stones found—Food of the people—Implements represent the Neolithic culture stage—The people were primitive hunters—The fauna—Scotch and other examples of shell-mounds—Darwin's description of people of Tierra del Fuego—Tumuli in Greece—Homeric heroes buried under tumuli—The same practice in other countries—Accounts in Homer—Herodotus on Scythian customs at burial of a chief—Julius Cæsar on Gaulish ditto—Tumuli, or barrows, in Denmark—Danish Sagas—Viking ship-barrows—Etruscan tombs resemble abodes of the living—Hut-urns—Origin of Architecture—Scandinavian 'passage-graves'—The Siberian 'yourt'—View of Professor Sven Nilsson, of Sweden, that underground dwellings were imitations of cave-dwellings—Burials in caves—Abraham—Canon Isaac Taylor on customs and worship of the Turanian race—Everything had its spirit (Animism)—British barrows—Researches of Sir R. C. Hoare, Stukeley, Dr.

# CONTENTS

Thurnam, General Pitt-Rivers, and Canon Greenwell—Thurnam's classification—The long barrows—Custom of the Iberians—Scottish Cairns—Mæshowe described—The 'Little People' or 'Fairies'—Runic inscriptions—New Grange (Ireland)—The bodies mostly unburnt—'The contracted position'—Absence of metal in chambered barrows (except gold)—The skulls long—'Ossuaries'—Canon Greenwell's explanation of the broken skulls—Cremation the rule in the Bronze Age . . . . . . . . . . . 191

## CHAPTER X

### THE 'LITTLE FOLK,' OR FAIRIES AND MERMEN

There is now a Science of Fairy Tales—Mythology and Folk-lore studied scientifically—Archæology confirms many of the old legends—The Scandinavian Sagas written in the Iron Age—Arrows of the Dwarfs, made of stone—Their supposed magical properties—Lapp skulls found in Sweden and Denmark—Evidence of Etymology—The dwarfs of the Sagas *not* mere inventions—Exaggeration is sure to arise when one race describes another—Esquimaux description of Englishmen—The dwarfs conquered by the Gothic people—Intermarriage took place—Dwarfs sometimes borrowed vessels or utensils—Jotnar, or giants—The Anakim of Canaan—Roman reports on the Germani—The word Fox-glove (Folk's glove)—The Venusburg in Tannhäuser—The Pied Piper of Hamelin—Clunie Castle in Perthshire—The Brugh of Boyne—The Tuatha de Danann—The Druids' decision between the Gaels (Milesians) and the Dananns—Sidhfir—St. Patrick and the Feni—Russian belief in 'Tshuds,' or former supernatural inhabitants of their land—Finns in the Isle of Sylt off the Schleswig coast—'Pucks'—Testimony of the late Mr. J. J. Campbell, of Islay—Santa Claus a Lapp—Story of Child Roland—Mr. G. L. Gomme on terrace-cultivation by pre-Aryans — Milton's 'Comus'—Picts and Finns—Picts' houses—Tacitus—Pict's house (Erdhouse) at Kingussie (Inverness-shire)—Saga of Thorgils—Skinboats or kyaks—Mermen—Finns allied to Laplanders—Shetland stories—Dr. Robert Sinclair on Capture of Finnish Brides—Matthew Arnold's 'Forsaken Merman'—Finns came from Norway—The longheaded (Iberian) race—The dark Highlander—Dolmen-builders—Professor Boyd Dawkins on the Neolithic homestead—Neolithic clothing in Andalusian cave—Egyptian traditions about the origin of their civilisation—Important discovery by Professor Flinders Petrie of Neolithic race in Egypt, who fill a gap in Egyptian history after the Seventh Dynasty . . . . . . . . . . 214

## CHAPTER XI

### RUDE STONE MONUMENTS

Dolmens—Their connection with stone circles—Kit's Coty House—
French Dolmens (chambered barrows)—Captain M. Taylor on Indian
Dolmens—Distribution of Dolmens—Not Druidical—Theories of the
migrations of Dolmen builders—Worsaaes' Chronological Scheme of
Prehistoric time for Scandinavian—Cups and vessels in Dolmens—
Ornaments—Pure Copper—Engravings—Distribution of rude stone
monuments—Standing stones in Scripture—Pillar Stones—Jacob,
Moses, Joshua—Hare or Hoar Stones—Cat Stones, Tanist Stones,
&c.—The Coronation Stone in Westminster Abbey (Lia Fail)—
Menhirs in France—Alignments at Karnac—Major Godwin-Austen
on the Khasia hill tribes (North India), and their standing memorial
Stones—Story of the 'Nongtariang'—Omens—How these menhirs
are set up—Religion of the Khasis—Propitiation of spirits—Family
vaults—Form of words used at a sacrifice—Mr. A. L. Lewis on the
transport of large stone blocks by the Khasis—Stone circles—Used
as courts of justice by Gothic people—For places of worship by the
Kelts in Scotland—Chapter-houses—Number of stones in a stone
circle—Scotch circles; Sepulchral deposits in them—In Irish circles
also Scandinavian circles—Avebury monument described—Silbury
Hill—Callernish circle and avenue—Cairn inside—Stennis circle—
Ring of Brogar—Druidical theories—Stone circles not of Scandi
navian origin . . . . . . . . . . 241

## CHAPTER XII

### STONEHENGE NOT DRUIDICAL

Much written on Stonehenge—But still a mystery—Etymology—Avenue,
'Friar's Heel'—' Altar stone '—Solar observations—Two barrows
inside—Description—The number of stones in outer circle here and
in other examples—The foreign stones—Where they came from not
known—' Slaughter Stone '—Mounds around Stonehenge, containing
chips of the same stones—The Cursus not a race-course Petrie on
evidence of incompleteness—The word 'Temple' seems inappropriate
—Transport of big blocks—Idea of Professor Petrie and Mr. C. H.
Read carried out in Plate X.—Theories about Stonehenge—Colonel
Forbes Leslie on Temples in India—Mr. W. G. Palgrave on similar
monuments in Arabia—Opinions of Mr. John Henry Parker and
Professor Nilsson—Mr. E. H. Palmer found stone circles with inter-
ments near Mount Sinai—Was Stonehenge erected by the dwarf
Fairy Folk?—Observations and arguments . . . . . 268

INDEX . . . . . . . . . . . 287

# LIST OF PLATES

| PLATE | | |
|---|---|---|
| I. | AN EVICTION SCENE AT WOOKEY HOLE, NEAR WELLS—OLDER STONE AGE . . . . . | *Frontispiece* |
| II. | HUNTING THE MAMMOTH IN SOUTHERN FRANCE . | *to face p.* 33 |
| III. | HUNTING THE REINDEER IN SOUTHERN FRANCE . | ,, 51 |
| IV. | HUNTERS FEASTING ON HORSE-FLESH IN SOUTHERN FRANCE . . . . . . . . . | ,, 83 |
| V. | HUNTERS IN ROCK-SHELTER AT NIGHT IN SOUTHERN FRANCE . . . . . . . . . | ,, 125 |
| VI. | SWISS LAKE-DWELLERS: AN EVENING SCENE—BRONZE AGE . . . . . . . . | ,, 173 |
| VII. | AN INTERMENT IN A LONG BARROW—LATER STONE AGE . . . . . . . . . | ,, 207 |
| VIII. | THE WARRIOR'S COURTSHIP, DENMARK—BRONZE AGE . . . . . . . . . | ,, 213 |
| IX. | TWO BRITISH WARRIORS—BRONZE AGE . . . | ,, 235 |
| X. | THE CONSTRUCTION OF STONEHENGE . . . . | ,, 241 |

# PREHISTORIC MAN AND BEAST

## INTRODUCTION

> When animals first crept forth from the newly-formed earth, a dumb and filthy herd, they fought for acorns and lurking places with their nails and fists, then with clubs, and at last with arms, which, taught by experience, they had forged. They then invented names for things, and words to express their thoughts; after which they began to desist from war, to fortify cities, and enact laws.—Horace, *Satires*, i. 3, 99.

IT is always a difficult matter to interpret the true meaning of those symbols by means of which the ancients chose partly to reveal, and yet partly to hide, their ideas with regard to the deep mysteries of Nature and of Life. But no one who studies the thoughts of old philosophers, whether from Egypt or Babylon, India or Greece, can fail to be greatly impressed, not only with their knowledge (which was wonderful for their age), but still more with the keen insight which, by contemplation and reasoning, they gained into the laws of Nature, and her workings in the phenomena by which man, from the cradle to the grave, is surrounded.

Sometimes one finds, as it were, a whole philosophy wrapped up in a symbol. Of one particular symbol we shall say more presently. Probably it will be news to many—and especially to that large class of persons who unfortunately read magazines and newspapers rather than books—to learn that not a few of the great discoveries of modern science, and of the conceptions which we are accustomed to associate with the great names of the present century, were anticipated by the philosophers of old time. Yet such is undoubtedly the case. 'Coming events cast their shadows before them,' and the old saying that 'there is nothing new under the sun' is constantly being verified. Lucretius, with his infinity of raining atoms, was the forerunner of Newton and Clerk Maxwell. In the days of Job, if ever such a person lived, there were observers of geological phenomena, and theories of the solid land being made by the agency of water. 'Speak to the earth, and it shall teach thee,' says the book that bears his name, and the writer was evidently aware that 'the mountain falling away cometh to nought, and the rock is removed out of its place: the waters wear away the stones; the overflowings thereof wash away the dust of the earth' (xiv. 18, 19). Egyptian and Babylonian sages, long before the Christian era, had taught their disciples that the earth had once given birth to some monstrous animals (such as the writer has elsewhere described).

In that old doctrine we can perceive the germ of the science of Palæontology, or the study of ancient forms of life. These men evidently had glimpses of the truth expressed in the following lines :—

> The earth hath gathered to her breast again
> And yet again the millions that were born
> Of her unnumbered, unremembered tribes.

Even the astronomers, unprovided with telescopes or any instruments of precision, and clouded as their minds were with the superstitions of astrology, yet believed in vast cycles of time, wherein the stars returned to their places after a circle of constant change. We must therefore give them credit for having discovered that wonderful movement of the earth's axis known as 'the precession of the equinoxes'—a process which it takes 25,000 years to complete!

Even the famous 'nebular hypothesis' of Laplace with regard

to the origin of the sun and planets had its prototype in Egyptian and Greek mythology. According to one of the old myths, the development of the world from chaos took place somewhat after the manner of a bird from an egg. When the first germ, or chaotic mass, had been made by a self-dependent Being, it required the mysterious functions of a second lower deity to make the mundane egg from which the whole organised world was produced. Thus, in that delightful play of Aristophanes, 'The Birds,' the chorus solemnly sing of an old doctrine telling how 'sable-plumaged night conceived in the boundless bosom of Erebus, and laid an egg, from which, in the revolution of ages, sprang Love, resplendent with golden pinions. Love fecundated the dark-winged chaos, and gave origin to the races of birds.'

One of the symbols of the ancients which seems to the writer particularly happy and suggestive is that old one of the Tree of Life, seen often on Egyptian tablets and temple-walls; trees were worshipped ages ago by prehistoric man as tokens of the power manifested by the Creator of Life and of the Universe. And it is possible that, as time went on, some of the old philosophers and naturalists of Greece began to perceive a deeper meaning in this symbol, and to realise that all the various forms of plant and animal life are, or have been, vitally connected together; that the same sap has nourished them, and that they are, as it were, but branches, some small and some large, of the great Tree of Life. When the naturalist of the present day turns his thoughts towards that greatest of all mysteries, the origin of Life, he finds that even in this dim region of speculation the ideas of Lamarck and of Darwin have been foreshadowed; for there is no doubt that several Greek and other philosophers had in their way got hold of the idea of evolution! Here, then, we have but one form of belief in the theory of the continuity of Nature, which the late Professor Tyndall expounded in his famous address to the British Association about twenty years ago.

According to Aristotle, Nature proceeds constantly, by the aid of gradual transitions, from the most imperfect to the most perfect, while the numerous analogies which we find in the various parts of the animal scale show that all is governed by the same laws. The lowest stage is the inorganic, and this passes into the organic by direct metamorphosis, matter being transformed into life. The greatest naturalist the world has ever seen was therefore distinctly an evolutionist, though not acquainted with the discoveries of Darwin, or with the great mass of evidence derived from the modern science of Palæontology. Nor is that all. He considered the theory of natural selection, or survival of the fittest, and rejected it as insufficient to account for the facts. And it is very striking to notice how, of late years, many even of the followers of Darwin are beginning to perceive that this cause, important as it certainly is, has been given too prominent a place in modern speculation. Darwin himself, in his later days, felt that he had gone too far. All who study life, whether the life on our planet to-day or the life of ancient geological periods, are in great expectation of the present discovery of some new law in evolution, some fresh *modus operandi* by which species undergo modification. Whenever such a discovery is made, it will be hailed with delight. The theory of Darwin and Wallace has undoubtedly marked an important epoch in human knowledge ; yet who knows but what another discovery may, ere long, be made which shall be received with even greater joy, and lead us still nearer to the desired goal ?

Aristotle believed that a ' Principle of Perfection ' operates all through nature, a theory which modern naturalists would do well to ponder over more seriously than they have done so far. This great conception may perhaps be regarded as partly a foreshadowing of that ' Law of Progress ' by which, as the geological record tells us, higher types of life from time to time appeared on the scene.[1]

[1] Anaximander (B.C. 610) taught that men in the beginning were engendered in fish, and, on becoming able to shift for themselves, were cast out and took to the land.

It is not too much to say that the whole philosophy of modern evolution, as applied to plant and animal life, is contained in that one symbol of the Tree of Life. But it is only in these latter days, after nearly a century of careful study of the 'Record of the Rocks,' otherwise known as the science of geology, and of the remains of animals sealed up in the strata beneath our feet (Palæontology), that we begin adequately to realise how true and suggestive it is. For, indeed, every tribe of plants or animals is, as it were, but one branch of that tree which has flourished for so many untold ages. Its age in years it is beyond the power of any man to reckon. The fossils by means of which the geologist determines the age of a stratum, and assigns it to one of the recognised periods, such as Cambrian, Cretaceous, or Tertiary, are but branches of that tree. From time to time a branch has dropped off, and become extinct; for example, the great Dinosaurian order of reptiles, while another, such as the Crocodilian order, has lived on with but little change to the present day.

We have a suggestion to make here. Perhaps the dinosaurs in time took entirely to the land, and some offshoots of theirs became mammals, so the struggle would be transferred to the land, hence crocodiles lived on in safety in the rivers. No attempt will be made here to deal with modern speculations regarding the origin of man from a lower type, and his relationship with the higher apes, such as the gorilla and the chimpanzee. Suffice it to say that he bears in his body the signs of his lowly origin; and that nearly all thinking people of the present day fail to see anything, in such a belief contrary to true theology, or to consider that there is anything degrading in the idea. Each one of us, before our birth, has gone through stages of existence exactly similar to those undergone by some of the higher mammals, and we then bore traces of the gills of a fish ! The modern view only teaches that the race progressed through the same animal stages as those represented in the pre-natal history of the individual.

After what has been said, perhaps the reader may not be

surprised to hear that, in the opinion of some students of ancient systems of thought, the account of the creation in the opening chapters of Genesis *implies* evolution. We find there a series of pictures in which the work of the Creator is briefly summed up It must always be borne in mind that the older Scriptures were intended to be amplified and expanded by the teacher in whose hands they were placed. People who wrote on religion before the days of printing never anticipated a time when their works would be in the hands of everyone, however ignorant or unlearned. They addressed themselves to those who, in the course of their education in the sacred colleges of Egypt or elsewhere, had been initiated into the inner mysteries of religion. Hence much was left out. The details could be filled in by the priests who taught the people. Bearing this in mind, it is not difficult to perceive that the account in Genesis, which agrees very closely with the Babylonian record on clay tablets, was written with the object of indicating briefly, and in a vivid manner, *some of the leading stages in a long series of operations.* The old view of a number of separate, disconnected, and sudden acts of creation is quite untenable. It would never have been held for a moment had our English divines grasped the idea, now dawning on the mind of our age, that the various writers who contributed to that collection of sacred books which we call 'The Bible,' in dealing with the deep mysteries of life or of the soul, wrote as mystics, and addressed themselves to such. For, indeed, religion can only be understood from the standpoint of the mystic.

What a world of controversy would have been avoided had this simple but vital truth been apprehended before ! That it was held by some of the Early Fathers is proved by their writings, as, for example, those of Origen. The ancient seers and teachers were not so childish or so ignorant of nature as to suppose that the world had been made otherwise than by a continuous series of slow and sublime operations ; although the form in which, for the sake of brevity and clearness, the narrative is cast might appear on the surface to bear a different interpretation. No, they only attempted to depict certain great significant stages in the work of the Creator, and the word 'day' may be understood to mean any period of time.

It is to be regretted that the revisers have so frequently placed in the margin the better translations. In Genesis i. 1, where we read 'the Spirit of God moved upon the face of the waters,' the marginal rendering 'was brooding upon' is much more suggestive of slow and continuous action, and the old Greek myth of the mundane egg was probably borrowed from the East.

The order of events all through is from the lower and more general to the higher and more special. First, all is darkness; then comes light, then the heavenly firmament, then the separation of land and water, rather suggestive of the geological truth that the land is made in the water (by sedimentation). Grass, herbs, and trees precede the appearance of cattle, and marine animals come before the terrestrial, which also is in accordance with modern science. All these stages, then, each higher than the last, are written down as coming before the appearance of man. Should we not, therefore, conclude that the lower animals were regarded as the *necessary* precursors of human beings, or, in other words, that the creation of Adam (the earth-man formed out of the dust of the ground, and typical of our lower bodily nature) was but the last of a grand, majestic, and orderly series of creative operations *all of the same kind?*

Eve, according to this interpretation, is the spirit, or divine soul of man, only realised when our lower and earthly nature is asleep or in subjection.

The old poets and writers would fain have us believe that the last word has been said about our ancestors. Brave men, they tell us, lived before Agamemnon, but such must ever remain 'unwept, unsung.' This is now no longer true. Where history or poetry fail, archæology steps in to fill up the gap, and holds a torch to light up our path in the dim recesses of the past, bidding us be of good courage, for rich fields of knowledge have been discovered, such as promise a full harvest in time. We need, therefore, no longer lament with Horace—

> Vixere fortes ante Agamemnona
> Multi: sed omnes illacrimabiles
> Urgentur, ignotique longa
> Nocte, carent quia vate sacro.

On the other hand, some persons appear to attach very little value to the records of written history, and quote the saying of a great man that it is a tissue of lies. What the opinion of such people would be of a work dealing with man before history we dare not surmise! But, whatever they may think, prehistoric man is far from being an unknown quantity. Through the labours of archæologists in various parts of the world, the present generation has gained some knowledge of his condition in life, his accomplishments, even of his thoughts.

There are two kinds of history, one written by men themselves, some of whom wrote at the time when the events they record happened, while others wrote long afterwards. This kind of history, useful as it certainly is, yet partakes of the imperfections of our human nature. *Humanum est errare.* The other unwritten kind may be said to have written itself in a sort of automatic machine fashion. Just as the history of the world in former distant ages is preserved in the strata beneath our feet, so that geologists can read it and tell us what the 'Record of the Rocks' says, so the early history of man is to be found sealed up in various accumulations that have in the course of time been formed—layers of earth in caves, of sand and gravel formed by rivers, heaps of refuse, like those of the rock-shelters of the Périgord, or the 'kitchen-middens' of Denmark, or the mounds (barrows) where men of earlier races lie buried. Then, coming to the borderland of history, we have the pyramids and tombs of Egypt, the buried cities of Chaldea or of Asia Minor, and the ruined palaces of

## INTRODUCTION

ancient Mexico. A goodly heritage has been bequeathed to the nineteenth and twentieth centuries. The record is there, if only it can be properly read. Already much has been read. We have, then, only to pull the lever of Mr. Wells's 'time machine' to find ourselves travelling backwards into the past at any velocity we please.[1]

But the machine in the present case must be the organ by means of which human beings reason—namely, the brain. That organ, if properly set to work, will guide us backwards into prehistoric times, enabling us to make the acquaintance of our primæval ancestors, to bring them back to life, so that they appear to live before us in the realm of imagination. Just as the science of palæontology teaches us how to 'restore' extinct monsters, and make their fossilised bones live once more, so the sister science of archæology should be made to furnish the necessary materials for 'restoring' primæval man in order that he too may be brought before us. Surely, so desirable an object is worthy of mental effort.

> The idea of his life shall sweetly creep
> Into your study of imagination,
> And every lovely organ of his life
> Shall come apparell'd in more precious habit,
> More moving-delicate and full of life,
> Into the eye and prospect of your soul,
> Than when he lived.[2]

Now, the facts on which the story of man is founded are not only very considerable in number, but of great interest and importance. Fortunately, there are still not a few primitive races

---

[1] It is curious to note that the author of this clever book describes a future state of society in which a conquered dwarf race, cunning in all kind of handiwork, lives underground, and so reproduces—probably quite unconsciously—a phase in human history which appears to have actually taken place in Europe! As will be shown in Chapters IX. and X., the Keltic people of Britain have traditions of such a race (elves, dwarfs, &c.), who were good builders and busy workers. See Mr. Jacobs' *English Fairy Tales*.

[2] *Much Ado about Nothing*, Act IV. Sc. 1, with slight alteration.

left whose manners and customs throw light on the ways of prehistoric man, and it is sincerely to be hoped that they will not be improved off the face of the earth before we have learned all they can teach us about the past. For it is not too much to say that savages have afforded clues without which archæologists would have been hopelessly at sea in their interpretations of facts.

And, conversely, the past throws light on the present. It is thus possible, in some degree at least, to trace things to their origins or beginnings. The knowledge by which we can trace back something to its first beginning always has a special charm, and in this way common things are often found to possess an unexpected interest. Everywhere there has been evolution: in society, in art, in morals, in religion, as well as in the vegetable and animal kingdoms. The world itself, nay more, the whole solar system, is the result of gradual unfolding or development.

It is believed that the first efforts in architecture were directed to the construction of chambers and passages in tombs, large upright stones being placed together inside a mound to make walls for a tomb, and other stones placed across them to form a roof. Hence, the three stones often seen in cromlechs and dolmens, of which the great trilithons at Stonehenge are survivals.

The graphic arts would appear to have grown up out of the attempt to decorate weapons, implements, and utensils, first with simple patterns, and then with outlines of natural objects.

Geologic time is divided into periods founded on the different groups of strata, their order of superposition, and the peculiar fossils contained in each, for each period had its ruling types. Historic time is likewise divided off into periods, marked by wars and conquests, and by the reigns of kings and queens. The reader will, therefore, naturally ask whether prehistoric time can also be marked off into more or less definite periods. Yes, there *are* prehistoric periods; in Europe, perhaps also in Asia, it has been found possible to create three fairly well-defined periods.

The nature of the weapons used by our ancestors affords the basis of such a division. First of all, stone was used for this purpose, but not exclusively, for it was supplemented by bone, &c. Hence the Stone Age is the earliest. Then, a long time after, came the invention of processes for smelting copper and tin in order to make bronze. That marks a very great advance in human progress, and we arrive at the Bronze Age. Later on iron was discovered, and the Iron Age began. But it is a mistake to suppose that this system, convenient as it may be for Europe, can be applied to all parts of the world, and when pushed too far it certainly has a harmful influence. Rather should we speak of a Bronze or Iron *State of Culture*, or Culture-Stage. (See p. 76.)

# Part I

# THE MEN
## OF THE
# OLDER STONE AGE

# CHAPTER I

### OUR EARLY ANCESTORS

*'Tis opportune to look back upon old times and to contemplate our forefathers.—Sir Thomas Browne.*

STARTING from the evolutionist's standpoint, and regarding man as an offshoot from some at present unknown branch of the tree of life, we proceed to the task before us, which is to endeavour to bring back prehistoric man from the dim vista of the past (where he has too long been lying in the darkness of oblivion), and to tell of his manner of life, not from any recorded words, because he could not write, but from his *deeds* as registered by solid and sound facts. In what words he expressed his varying thoughts will never be known; but, fortunately, we have other expressions of his. Every deed done, every weapon or utensil made, every ornament designed, is but an expression of thought. And hence, if we can but interpret aright his implements, utensils, &c., his tombs, mounds, monuments, and rock-shelters, we shall find ourselves in possession of a most valuable and interesting record, more safe, in some ways, than any written documents. Here nothing is extenuated, nor aught set down in malice.

Perhaps one of the first inquiries which suggests itself with regard to the origin of man is the question, 'Where did he first appear?' At present there is, unfortunately, not enough evidence to settle this interesting problem; but we may state in passing that our greatest naturalists, such as Darwin, Huxley, and Wallace, have given reasons for believing that the old world, and not the new, was the scene of his first appearance on earth, possibly in the period known to geologists as the Miocene. The Java skull found by Dr. Dubois, of which we have heard so much lately, is described in Chapter VII. Its geological age is not yet known with certainty. These great naturalists have pointed out that the gorilla and the chimpanzee, his nearest relatives, are not to be found in either of the two Americas, Australia, or the Pacific Isles, but are distinctly old-world forms. But, while some think that Africa was the land of his birth, others consider that more probably it was Asia. In the present state of our knowledge, all that can be safely said is that, since Palæontology shows the higher forms of mammalian life to have originated in the land mass of the northern hemisphere, it would seem most likely that man also arose there. It must not be forgotten, however, as Sir Henry Howorth points out, that human skulls were found by Mr. Lund in South American caves, associated with extinct mammalia.

With regard to the origin of civilised man, as opposed to savages, the remarkable fact has frequently been noticed that all high civilisation has developed in or near the north temperate zone, where the natives appear to have long ago arrived at a higher state of civilisation than those of either the frigid or torrid zones. It has therefore been suggested that an extensive land area, combined with a *temperate* climate, is a condition favourable to the development of intelligence, and perhaps the necessity of storing up food for the winter may have tended to develop those latent faculties which must certainly have existed in early races, even as they do now in the natives of Africa.

The earlier history of the human race is naturally the most difficult of interpretation; and the first period of which we have any knowledge at all is that of the Older Stone Age, which Sir John Lubbock calls the Palæolithic period. At this stage of our investigation it will be necessary to call in the aid of the geologist and the naturalist to help us in our interpretation of such records as have survived. These are plateau and river-gravels with their buried flint implements; bones of men and beasts, &c.; terraces, glacial moraines, caves and rock-shelters, with their old buried floors and hearths, implements of bone, and other things.

Lucretius truly remarks that 'arms of old were hands, nails and teeth, and stones, and boughs broken off from the forests, and flame and fire, as soon as they had become known. Afterwards, the force of iron and copper was discovered ; and the use of copper was known before that of iron, as its nature is easier to work, and it is found in greater quantity' ('De Rerum Natura'); but he also conceived of a time when man was ignorant, both of the use of fire and of skins for clothing his body, as the following passage will show :

Necdum res igni scibant tractare neque uti
Pellibus et spoliis corpus vestire ferarum,
Sed nemora atque cavos monteis sylvasque colebant,
Et frutices inter condebant squalida membra,
Verbera ventorum vitare imbreisque coacti.

*De Rerum Natura*, v. 951.

About the close of last century, or a little later, certain geologists had arrived at the conclusion that some of the human remains discovered embedded in the floors of caves were as old as the extinct mammalia whose bones were found in the same layers. But in those days scarcely anyone dared to attribute to the human race an antiquity greater than about four thousand years, which was in accordance with the calculations of the pious Archbishop Ussher, whose chronology remained for so long unquestioned, since it was vainly supposed to be based on Scripture. Anyone, therefore, who dared to assert that man was contemporary with the extinct animals of a former geological period was sure to be denounced. But truth was bound to triumph in the end ; and in the year 1841 a zealous French antiquarian, M. Boucher de Perthes, observed in ancient river-drift at Abbeville the bones of extinct mammals associated in such a manner with flint implements of a very rude type as to clearly demonstrate that both belonged to the same period. And although, in a few years, many more such stone weapons were found in these old gravels and loams, yet their discoverer was regarded both in England and in France as an enthusiast, almost as a madman! It was not till 1859 that attention was at last directed to these most important discoveries by Dr. Falconer, and Prestwich, Evans, Lyell, and Lubbock. It is also somewhat strange that, long before these discoveries in the valley of the Somme attracted so much attention, discoveries of like significance had been made in England, in the years 1715, 1800, and 1836, but they had been entirely forgotten. M. Boucher de Perthes had also been anticipated in Belgium, where, to his honour be it said, the late Dr. Schmerling, of Liège, had shown that man and certain extinct mammalia were contemporaneous. Such is the fate of what Lyell calls 'unwelcome intelligence, opposed to the prepossessions of the scientific as well as of the unscientific public.'[1]

At about the same time the Rev. J. MacEnery, a Roman Catholic priest, arrived at the same conclusion from discoveries made in the now famous bone-cave known as Kent's Cavern, Torquay, but did not survive to publish the results of his investigations.

The implements found in these old high-level river-gravels, far above the present course of the stream, are all of a primitive kind; nevertheless, they are capable of being roughly classified. Some are mere chipped flakes, which were probably used as knives; others may have been arrowheads, though this is doubtful; some are pointed weapons, suggestive of lance or spear heads; some are axes or hatchets, and others are oval instruments with a sharp edge all round, but very large, shaped, as the French expressively put it, like cats' tongues. The researches of Mr. Worthington Smith have also revealed small scrapers for preparing skins, hammer-stones used in breaking off flakes, anvil-stones for a like purpose, punches, borers, &c. They generally are

[1] *Antiquity of Man*, 4th edit., p. 71, and *Prehistoric Europe* (1881), by Prof. James Geikie, p. 4.

stained of the same ochreous colour as the flint of the gravel in which they occur, and often found at depths of from fifteen to twenty feet from the surface, facts which speak in no uncertain tone of their great antiquity. Their edges also are considerably worn. It has been suggested that some of the larger pointed implements [1] may have been fastened into the ends of poles for digging; but perhaps the majority of them were only intended to be held in the hand. Some savages of the present day use simple flint or stone knives inserted in wood, or simply covered at one end with bitumen, or skin of some kind, to form a handle or grip.

None of the implements found in these old deposits have their edges ground, but only chipped; this is an important distinction, which separates them from those found at lower levels, which show a great deal of finish, some of them being carefully ground and polished. Such belong to the Neolithic or Later Stone Age, while the former are known as Palæolithic. But for all that, Neolithic implements are often unground.

When we call in the geologist to interpret to us the meaning of the older gravels, he tells us that they were deposited a very long time ago by river action, at a time when the physical geography of the region of the Thames or the Somme was not what it is now. The sides of their valleys are strewn with patches of gravel left there from time to time as the river slowly cut its way down to lower levels. Thus we have evidence that since Palæolithic times these rivers have excavated their valleys to the extent of from 60 to 100 feet. Sir Henry Howorth offers a totally different explanation. But we cannot stay to discuss this wide subject here (see Chapter IV.).

Judged by every standard, either geological or anthropological,

---

[1] The long oval and pointed implements (seen in the British Museum Prehistoric Saloon Cases, 51-60) are difficult to explain. The writer believes them to be prototypes of the axe. The Neolithic axes appear to him to be often based on this type.

these implements from the river-drifts or gravels certainly bear witness to a high antiquity, though nothing like so great as those maintain who have allowed themselves to be carried away by the glamour of glacial theories, based on unwarranted astronomical assumptions. We shall deal with this question in other chapters on 'The Myth of the Great Ice-Sheet' and 'The Antiquity of Man,' only remarking meanwhile that, as far as we can see, neither science nor common-sense give any warrant for attributing to the men of the Older Stone Age, or Palæolithic period, such an antiquity as the 80,000 years or more with which the late Mr. Croll and others misled the public, who, though ignorant for the most part of science, are ever ready to listen to something new and startling, and the more startling it is the better they are pleased.

Not but what we believe man to have a very high antiquity; that, however, is quite a different matter from saying that such antiquity *has been already proved*. It is most important to make very clear distinctions between conclusions which are only probable (and may take a long time to work out) and those which are actually demonstrated. We ourselves hold that nothing has yet been discovered to trace back the footsteps of the human race more than about 15,000 to 25,000 years.

But instead of attempting to calculate in years the precise antiquity of these times, it is better—since geology has no mathematical basis—to try and form some conception of the ages that have elapsed by studying the great changes that have since taken place on the face of the land (see Chapters VI. and VII.). Physical geography, properly interpreted, tells us that at this time the continent of Europe stood at a higher level than it does now. The whole of the North Sea, even between Scotland and Denmark, is not more than fifty fathoms, or 300 feet deep, while the Irish Sea does not go beyond sixty fathoms; and there can be no doubt that at this period the British Isles were not only joined together, but formed part of the mainland, and were in touch with France and Germany.

According to some geologists, our own Thames, and other

rivers on that side of England, were then but tributaries of one large river which flowed northward through this continent. The position of this ancient river is said to be still more or less clearly indicated by soundings marked on charts of the North Sea. On the eastern side, the present Rhine formed part of another branch or tributary, and the river may have emptied itself into an Arctic Sea somewhere between the Faroe Islands and Iceland. The western margin of our continent at this time is indicated by the hundred-fathom line which is shown on the charts at a distance of some 200 miles from Ireland and Scotland. Looking south, we find that Spain and Africa were united, while the Canaries and the Azores formed part of the mainland, and the waters of the sea *may* have flowed over what is now the Great Desert of Sahara. Sir Henry Howorth, however, informs us that the French geologists have shown the Sahara to be simply an old land surface with Palæolithic flints.

Speaking of the changes that have taken place in the South of England, Sir John Evans says : 'Who, standing on the edge of the lofty cliff at Bournemouth, and gazing over the wide expanse of waters between the present shore and a line connecting the Needles on one hand and the Ballard-Down on the other, can fully comprehend how immensely remote was the epoch when what is now that vast bay was high and dry land, and a long range of chalk downs, 600 feet above the sea, bounded the horizon on the south? And yet this must have been the sight that met the eyes of those primeval men who frequented that ancient river, which buried their handiworks in gravels that now cap the cliffs, and of the course of which so strange but indubitable a memorial subsists in what has now become the Solent Sea.'

The geology of the Pleistocene period is full of controversies with regard to certain highly interesting but difficult problems, such as the nature and origin of the peculiar 'drift deposits' of the South of England, below the Thames, and their true relation to what is called the boulder clay. The latter in its turn has, perhaps, given rise to more discussion than any other subject in the whole science of geology (see Chap. IV.). It is a deposit of the Glacial period, formed partly by the action of ice, but a large part of it may represent a pre-Glacial soil. There is another unsolved problem. On returning to the subject of our present chapter, we find this question facing us, 'When were the higher Palæolithic river-gravels formed?' The literature of all these questions is truly enormous, and always highly detailed and technical. It is, therefore, impossible to discuss such subjects in a book like the present. To endeavour to do so would involve laying before the reader a mass of detailed evidence from sections in gravel pits, &c., such as would be almost unintelligible, with the probable result that he would lay the book down at once.

There is also the difficult question of the origin of the plateau drifts, of which much might be said. All we can attempt to do is to indicate in passing the direction in which the evidence at present brought to light seems to lead. And here we may be permitted to remark that many years of careful research are necessary, before sufficient evidence can be collected to enable the geologist to form true judgments on these difficult problems. His position is, at present, like that of the jury in a long, difficult trial before all the facts on both sides have been elicited. But, with regard to the problem before us, perhaps we may be allowed to say, in spite of what is set down in many text-books, that of late years many have come round to the view held by that most distinguished of living geologists, Sir Joseph Prestwich, that the old river-gravels with Palæolithic implements were formed by the Thames and other rivers *during the latest phase of the Glacial period*. The 'Contorted Drift' seems to testify to the action of floating ice; and certainly the evidence from contained bones and teeth of extinct mammalia strengthens the idea. They do *not* appear, therefore, to be post-Glacial, as is often stated in books on geology, but we shall return to this subject in a subsequent chapter.[1]

Unfortunately, on account of the highly technical nature of the evidence, and other causes, we can only allude very briefly to a recent discovery of flint instruments of a very primitive character at high levels on the chalk plateau of Kent, which promises to be of the highest interest and importance. If the views of Sir Joseph Prestwich,[2] who has fully reported on the discovery, contain the true interpretation, they lead us back, at a single bound, to a period far more distant than belongs to any true discovery that has yet been made, and our ideas with regard to the antiquity of man must be enormously enlarged. But in a subject of this kind great caution is necessary, and the writer will not express any opinion. The implements in question were found by Mr. B. Harrison and others, mostly on the surface, but, in a few cases, they seem actually to belong to the peculiar clay drift capping the plateau. In 1894 some were found at a depth of from five to six feet below the ground ; but, at present, the geological evidence is not complete.[3]

We spoke just now of the opinion at present gaining ground

---

[1] The student may consult the following works :—*The Antiquity of Man*, Sir C. Lyell ; *Cave Hunting*, Prof. Boyd Dawkins ; *Early Man in Britain*, Prof. Boyd Dawkins ; *Prehistoric Times*, Sir John Lubbock ; *Prehistoric Europe*, Prof. James Geikie ; *The Great Ice Age*, Prof. James Geikie ; *The Mammoth and the Flood*, Sir H. Howorth ; *The Glacial Nightmare and the Flood*, Sir H. Howorth ; *Geological Magazine*, 1893-5, papers by Sir H. Howorth ; *Quarterly Journal Geological Society*.

[2] While correcting these pages for the press, we read with sorrow of the death of this great geologist, a man loved and revered by all who knew him.

[3] This subject can be more appropriately dealt with in the present chapter than in that devoted to the Antiquity of Man, because the question turns very much on the relation of this plateau-drift to the ordinary river-drift. For further information the reader is referred to the following :—*Quarterly Journal Geological Society*, 1889, p. 270, and 1891, p. 126 ; *Journal Anthropological Institute*, 1892, p. 246 ; Prestwich's *Collected Papers on Controversial Questions in Geology*, 1895 ; *Natural Science*, April and October, 1895.

that Palæolithic man lived about the end of the Glacial period; but must pause here to describe the above discovery, which suggests his being actually pre-Glacial!—a result hitherto regarded as almost 'beyond the dreams' of geological 'avarice!'

According to Sir Joseph Prestwich, the implements appear to have been carried down from south to north, with a drift unlike that of the neighbourhood, on to the plateau. Where did they come from? His interpretation at first appears a bold one; for he believes that they came from certain high grounds, or uplands, now no longer existing, in the centre of the Wealden district, their disappearance having been due to the subsequent 'denudation' of the Wealden area—that is, a slow wearing away, by the agency of 'rain and rivers,' in the manner familiar to every student of geology, even the youngest. These old uplands may have formed a low mountain range from two to three thousand feet in height! No wonder, then, that some who study these matters are inclined to doubt the truth of the late Professor's theories, while others await with eager expectation the establishment of conclusions put forward by so highly venerated an authority, who is everywhere recognised as the Nestor of English geology.

The implements are of very rude make, and of peculiar types; so simple, in fact, that some geologists and others refuse to look upon them as the work of men's hands, as do Sir John Evans and Sir Henry Howorth. But, surely, no candid person on looking at the few specimens figured by Prestwich in his recent book on 'Controversial Questions in Geology' could fail to perceive that they bear evident traces of selection and trimming with a definite purpose? No accidental breaking and bruising could possibly account for a succession of neat and definite chips all round the edges of a flake, such as we see here. They are deeply stained, and appear in every way to be of the same age as the drift itself; and, with very few exceptions, they are not met with in the valley-gravels below.[1]

[1] Supposing, however, that they should prove eventually *not* to belong to

They are mostly of small size, from two to three inches long, and, therefore, could easily have been held in the hand. Viewed as a whole, they seem to be as widely different from those of the Palæolithic Age as the latter are from weapons of the Neolithic or Later Stone Age, and that is saying a great deal, but rudeness of make alone is not always a sure sign of antiquity. It is only because combined here with other evidence that it may be regarded as important. Probably the implements were made for such purposes as breaking bones for the sake of the marrow, scraping skins, bones, or sticks. The large massive adzes of the river-gravels are but seldom met with here, and probably were brought to this region later. Some are split flint pebbles, chipped round the edges for cutting or scraping. Similarly made scrapers, according to Dr. Blackmore, of Salisbury,[1] are still in use among American Indians, who, when they want a scraper, select a pebble, which they split and then trim off the edges.

Undoubtedly the most important discovery of human remains at present ever made is that of the year 1886, at Spy, in Belgium. The bones, which belonged to a man and a woman, were found in a terrace giving access to a cave (not in the cave itself, as some accounts say), and embedded in a hard breccia containing fragments of ivory and many small flints *and arrow heads*. They appeared to be in a recumbent position and undisturbed. About this discovery there can be very little doubt. All the

the plateau drift, the writer would suggest that perhaps they might have been the work of a small race of prehistoric men, driven by the conquering occupants of the valleys to seek refuge on the higher grounds of the plateau, while their enemies enjoyed the greater warmth and fertility of the valley. This is a process that has happened many and many a time since in the history of nations. In Britain we find the Kelts in the mountains.

[1] Visitors to Salisbury should not fail to inspect the splendid and unique collection in the Blackmore Museum, where the whole subject of flint and stone weapons is fully illustrated by careful and wise arrangement, as well as by comparison with the weapons of modern savages in all parts of the world. Those who travel abroad should visit the famous Museums of Berlin (Völkerkunde), Copenhagen, Munich, and St. Germain's.

authorities agree that these remains belong to a period very far distant from the present—probably the Glacial period; this is proved by the fact that the bones of certain now extinct mammalia were found at the same level and immediately above. These belonged to the mammoth, tichorhine rhinoceros, cave bear, hyæna, reindeer, and other extinct forms.[1]

Speaking of the Spy skeletons, the late Professor Huxley said: 'The anatomical characters of the skeletons bear out conclusions which are not flattering to the appearance of the owners. They were short of stature but powerfully built, with strong, curiously curved thigh-bones, the lower ends of which are so fashioned that they must have walked with a bend at the knees. Their long, depressed skulls had very strong brow ridges; their lower jaws, of brutal depth and solidity, sloped away from the teeth downwards and backwards, in consequence of the absence of that specially characteristic feature of the higher type of man, the chin prominence. Thus these skulls are not only eminently "Neanderthaloid," but they supply the proof that the parts wanting in the original specimen harmonised in lowness of type with the original.'

Another passage by the same authority on the above subject is worth quoting, were it only for the gentle irony it contains: 'After all due limitations,' he says, 'they give us some, however dim, insight into the rate of evolution of the human species, and indicate that it has not taken place at a much faster or slower pace than that of other mammalia; and, if that is so, we are warranted in the supposition that the genus Homo, if not the species which the courtesy or the irony of naturalists has dubbed sapiens, was represented in Pliocene or even Miocene times. But I do not know by what osteological peculiarities it could be deter-

---

[1] The almost entire absence of human bones in old river-gravels seems at first sight rather remarkable. Lyell pointed out that the primitive inhabitants of the valleys of the Somme and Thames were probably too wary and sagacious to be often surprised and drowned by floods which might sweep away an incautious elephant or rhinoceros. This idea, however, is open to criticism, because these animals are known to be very wary. But even if some human beings were drowned, the chance of preservation was small.

mined whether the Pliocene or Miocene man was sufficiently sapient to speak or not ; and whether or not he answered to the definition "rational animal" in any higher sense than a dog or an ape does.'

The exact period of time when the Spy man lived is more or less a matter of conjecture, but for the present we may perhaps be justified in placing him somewhere about the period of the River-drift man now under consideration. One might be tempted, judging merely from the anatomical features, to put him further back in time, and to regard the bones as possible relics of the Pliocene period; but we are all on the look-out for Pliocene man, and for that very reason it is necessary to be very cautious, otherwise we may be prejudiced in favour of the idea, and the independent witness would at once reply that the wish was father to the thought. Of late years the geological world has been so frequently excited by reported discoveries of the remains of both Miocene and Pliocene man, all of which have, on fuller investigation, broken down, that one almost instinctively regards each new announcement as a false alarm (see Chap. VII.).

Unfortunately, river-gravels are not the most favourable material for the preservation of bones ; so that, although geologists have kept a sharp look-out for many years past, at present no skeleton or skull has been found in a river-gravel which can be pronounced with absolute certainty to be that of a Palæolithic man. But a recent discovery of human bones at Galley Hill, Northfleet, so nearly satisfies all the conditions required both by the anatomist and the geologist as to deserve some notice here.

The remains in question, consisting of a nearly complete skull and part of the lower jaw, with some of the grinders, two thigh bones, parts of a clavicle and humerus or arm bone, with fragments of a pelvis and sacrum, together with portions of ribs, all showing evidence of great antiquity, were discovered in 1888 in a high-level gravel [1] ninety feet above the Thames. Undoubted Palæolithic

[1] The bones appear to have been found in a silty clay-bed in the gravel, hence their preservation, clay being far less pervious to water than gravel.

flint implements occur in this same gravel, and bones of extinct mammalia of that period have been met with at a similar level not far off. The writer had the pleasure of hearing this find described, in a very carefully written paper read before the Geological Society in the present year, by Mr. E. T. Newton, palæontologist to the Geological Survey. The skull is very long and narrow (dolichocephalic), with strongly developed brow ridges, and the jaw was apparently projecting, as in modern savages. The bones were carefully compared with the well-known specimens from Spy, Neanderthal, and Naulette, which are generally accepted as of Palæolithic Age, and in several important features they agree.

A keen discussion followed the reading of this paper, in which such authorities as Sir John Evans and Professor Boyd Dawkins took part. Much difference of opinion was expressed; the two latter authorities were unfortunately not convinced by the evidence brought forward, and preferred to suspend their judgment. Many others, however, expressed the opinion that the remains really belonged to a man of the River-drift period, a view shared by the present writer. Those who were unwilling to entertain the opinion of Mr. Elliot, who discovered the bones, were inclined to think that this was a case of interment in the Later Stone Age, or Neolithic period.

The controversy thus set up turns almost entirely on this one point, but we can see no reason to doubt the statements of those who reported the discovery (also of the workman who found the skull), that the gravel and loam at this spot showed not the slightest evidence of having been disturbed. And therefore, until the other side can show that they *were* disturbed, we shall maintain that the Galley Hill find exhibits what has been so long awaited with keen expectation by geologists and anthropologists, namely, a specimen of the mortal remains of an individual of that very ancient and lost race of men who lived in Britain with the mammoth and woolly rhinoceros, the cave-bear, the cave-lion, hyæna, and other creatures. These flourished during the Glacial period, when the aspect of the country was very different from what it is now, and when there was no English Channel.[1]

Next in importance to the Galley Hill discovery comes a find by Mr. Worthington Smith. This eager collector of flint implements states that the only undoubted relic of the kind which he has come across is a fragment of a skull found in Suffolk, in seven and a half feet of 'brick-earth' or loam, in the parish of Westley, found by Mr. H. Prigg. The discoverer says there can be no possible doubt as to the great antiquity of the fragment, and that the red loam in which it was found was deposited by a river ages before the excavation of the Linnet to the south. He discovered it in what is called a 'pocket' eroded in the chalk. In some of the adjoining 'pockets' teeth of the mammoth were found and some implements of Palæolithic type. At the time of the discovery a series of such small pits were being worked by the men, one of whom declared that thirty years previously an entire skeleton of a man was discovered in one of them, at a depth of about eight feet from the surface! It is a thousand pities that such a precious find should have been lost for want of someone who would know its value to the scientific world; however, this only tends to confirm the Galley Hill discovery, and we may confidently expect that in time, as more people are interested in these matters, other and still more satisfactory discoveries will serve to enlighten us about our Palæolithic ancestors.

Of late years Mr. Worthington Smith has, by his indefatigable labours in collecting and studying flint implements, added some

[1] It should also be borne in mind that the depth (eight feet) at which the bones occurred is decidedly against the idea of a Neolithic interment. Although we can hardly say it is decisive, there was no sign of a burial mound, or barrow, here (see Chap. IX.) such as mark the resting-places of chiefs and heroes. Nor is it at all likely that the body of a man of no distinction would be buried at such a depth below the surface. Sanitary considerations did not trouble the minds of primæval man.

valuable material to our at present small knowledge of the River-drift men and their doings. He and Mr. Spurrell have found some of the old sites where they lived, loved, feasted, made implements, dressed the skins of wild beasts caught in the chase, and probably spent their time very much in the same way as bushmen in Australia and other primitive races are doing at the present moment. Probably they even erected rude shelters or tents, wherein to seek protection from the elements, and lie down to sleep at night. These old sites he calls 'Palæolithic floors.'

It is hardly necessary to say that the sites are now buried up by later accumulations of sand, gravel, &c. One of these is at Caddington, in Hertfordshire, thirty miles north of London, near Luton. Here he has actually found the little heaps of flints which these primitive people collected from the neighbouring ground and brought together to be made into flint implements; some of these heaps were nine feet in diameter, others smaller. Around them were hundreds of 'flakes' produced in the manufacture of the implements, 'cores,' or blocks from which flakes were struck off, 'punches' with which flakes were made, and other objects. These implements are all of small size, compared to many which have been found in river-gravels (a point which may have some importance if we only knew the true interpretation of it), and pretty much of the same type as those of Moustier, in the Dordogne district. Does it mean that these people were of smaller stature, like the pygmies of Central Africa? Some of the small scrapers are so well made as not to be distinguishable from those of the Later Stone Age, or Neolithic times. Similar 'floors' are met with in North-east London, where a thin stratum represents it, and the implements are of the same style.[1]

Thus at Stoke Newington Common, at a height of ninety feet

[1] A portion of such a 'floor,' with the shoulder of a mammoth, shells, and flint implements, is to be seen in the British Museum (Anthropological Gallery).

above the sea, Mr. W. Smith discovered such a 'floor' buried up under four or five feet of sandy loam, and resting on the surface of the old Thames gravel. Here the geological record, as may easily be interpreted, gives us an interesting little detail with regard to the life of our ancestors here; namely, that these sites were liable to be occasionally flooded, for a thin layer of sand here separates two old land-surfaces, telling us that after the lower and older floor had been invaded by the river and covered with sand, our friends came back again, nothing daunted, and, squatting down on the new surface, made new settlements for themselves where the manufacture of flint weapons went on merrily as before, just as people come back to their own city after an earthquake.

A very important fact with regard to these old sites is that the brick-earth in which they occur is overlaid by what is known as the 'contorted drift,' in some cases disturbing it, or even forcing its way under it. Now no ordinary deposit formed by water ever shows contortions, so we have here a geological fact of no small importance, and one which throws useful light on the conditions of climate, &c., that prevailed during this period. Contorted gravels point to the presence of ice pushing along and disturbing in its course the strata of sand or gravel which had been peaceably deposited by a river in the ordinary way. 'The undulation and contortion of the upper drift material seem to show that it was laid down by moving ice or frozen mud from the north. The abraded and whitened implements and flakes, sometimes embedded at all angles in the upper contorted drift, were, no doubt, caught up from old exposed land surface and carried southwards by slowly-moving half-frozen mud' (Worthington Smith). Whatever is the exact interpretation, there can be little doubt of the former presence of floating ice here. Some geologists, however, take a different view. Sir Henry Howorth thinks that water in violent motion can make contortions, and writes: 'I cannot see any ice-work here at all.'

We have already remarked upon the small size of those

implements which are found in the old 'floors,' and it now appears that there is a kind of stratification or order of succession in the implements themselves; for in the underlying gravels occurs, according to the same authority, quite another series of implements, bearing the marks of a greater antiquity, such as the condition of the flint and the style of the implements themselves. They are generally of larger size and without the secondary chipping; scrapers occur among them; but at a still lower level. At a depth of thirty feet or more, yet another class of implements makes its appearance; these are of a very rude kind, much worn, and deeply stained with iron ore. Since scrapers and knives are absent here, we may perhaps, conclude that the savages who lived at this early period of our Thames Valley deposits were unacquainted with the art of dressing skins.

Although the contorted drift itself contains implements which are often much water-worn, Mr. W. G. Smith's researches so far go to prove that Palæolithic man here retired before the advancing drift. Still, we must remember that this conclusion rests only on what is called negative evidence, and so may not be final.

It is possible to glean something with regard to the manner in which the River-drift men set to work to make their implements, and readers of Mr. W. G. Smith's interesting work will find there ample descriptions of their tools, anvils, &c. It is not difficult to detect the core or nucleus from which flakes have been struck off. Numbers of such cores have been found by him on the old floors, and sometimes the flakes that were struck off from it have been found, and each one restored to its original place on the block of flint! Such a block may be seen in the Natural History Museum, Cromwell Road (Gallery 1, Geological Collection), in a case with old human skulls. Probably they were dislodged with 'hammerstones' either of quartzite or of flint (in a harder state), for flint when first extracted from a chalk quarry is softer than that which has been exposed to the air for some time.

Besides these, the anvil-stones have been found on which the

Plate II.

HUNTING THE MAMMOTH IN SOUTHERN FRANCE

core was placed while flakes were struck off from it; they are large in size, and can be recognised by a practised eye on account of the marks of hammering and 'bulbs of percussion.'

He has also found stakes, which they sharpened in order to drive them into the ground, and little stone beads used for necklaces, which are in reality small fossil sponges, with a hole formed naturally in the centre and known as cosinopora.

The following graphic picture of the animal life of this period is from the pen of Mr. Worthington Smith, in his recent work entitled 'Man, the Primæval Savage :'—

'The fossil bones found associated with the stone implements of primæval man show that the following animals, amongst many others, were man's companions. The hippopotamus, mammoth, elephant, rhinoceros, lion, wild cat, bear, hyæna, ox, bison, and wild horses, the latter perhaps then faintly striped like a zebra. The hippopotamus reached what is now the Thames by rivers and the seashore from Africa. Not being a flesh-eating animal, it would not be much dreaded by its human companions; the old bulls would, however, sometimes scatter human companies. Neither would the hairy mammoth and the straight-tusked elephant molest the men further than by an occasional charge from a furious old bull. The rhinoceros was doubtlessly a dangerous animal, and no man would dare to face it. Men, horses, oxen, deer would all give a wide berth to the different species of rhinoceros; doubtlessly, men, women, and children, as well as other animals, were often mauled, ripped, and killed by them; the stealthy and terrible lion, silent and swift of foot, together with the spiteful and ferocious wild cat, would always strike terror into the heart of the primæval savage. The bears would occasionally stray from their dens and attack wild horses, wild oxen, and deer. The formidable grisly bear would doubtlessly sometimes attack the human families. The cowardly and terrible hyæna would frequently chase or pounce on men, women, and children, as well as on wild horses and oxen; it would at times stealthily discover, bite, tear, and kill members of the human family at night. Packs of wolves, famished with hunger, would make short work of old men, women, and children. The fox lived in the woods, whilst the roe, the red deer, the reindeer, and the gigantic Irish elk would frequently be seen in glades and open spaces. An ape lived in the forest, moles, beavers, and otters frequented the rivers.[1]

'The interest in all other animals completely palls before the presence of man himself. Amongst all the other living creatures, what kind of man was the earliest human savage? Let us suppose that it is night, and that we have reached, under cover of darkness, a haunt of primæval savages. The nocturnal sounds are strange and startling; we hear the terrific snorting, blowing, and splashing of herds of hippopotami as they wade and walk through the water of the Thames and Lea, or crash through the bracken and bush of the river banks; we hear the trumpeting and bellowing of elephants, and the roaring, snorting, and grunting of the rhinoceros. The roar of the lion, the devilish howling cry of the wild cat and the hyæna, the growl of the bear, the roar of the bull, the howl of the wolf, the bellowing of the stag, the bark of the fox, the neighing of the wild horse, and the chattering of the ape are heard.'

The following rather fanciful picture of the 'men and women of the day' is interesting, but in our opinion hardly does them justice. Nor does it seem likely that men who could make tools and wore ornaments were denied the gift of speech. Some anthropologists consider the natives of Tasmania may be taken as living representatives of Palæolithic man. He continues thus:—

'But of all the sounds in which we are interested, none equals in interest and importance the voice of man himself. Man's voice was at that time probably not an articulate voice, but a jabber, a shout, a roar. A shriek or groan of pain is heard, a

---

[1] These animals were probably not all living together in any one district all the year round. Some preferred the heat of summer, others the cold of winter.

shout of alarm, or a roar of fury. Loud hilarious sounds as of strange laughing are heard, and quick, jabbering, threatening sounds of quarrelling. Coughing is heard, but no sound of fear, or hate, or love is expressed in articulate words.

'If we imagine the darkness to have lifted, we see the men and women standing about or crouching—many carrying bones and stone tools—near fires. There is one central fire and several minor fires bounding the fringe of the human haunt. The fires are kindled from sparks (derived from concussion of flints) applied to dry grass. Some of the men and women are feeding the flames with ferns, twigs, tree-branches, and logs. Other men and women are seen sitting or lying about in dens or hovels formed of tree-branches and stones, or resting under bushes, trees, fallen trunks, or natural sheltering banks of earth. Hairy children are seen running about or crawling about on all fours. Bones, some with putrid meat attached, are seen strewn about in all directions.

'The human creatures differ in aspect from the generality of men, women, and children of the present day; they are somewhat shorter in stature, bigger in belly, broader in back, and less upright. They have but little calf to the legs. The females are considerably shorter than the males, they bear children in their early youth, and cease to grow.

'All are naked [1] or only slightly protected with ill-dried skins. They are much more hairy than human creatures of the present time, especially the old males and children. In this character they resemble the present race of hairy Ainos of the northern islands of the Japan Archipelago.[2] The hair is long and straight, not curly, the colour probably bright chestnut red, and the skin copper-colour. The heads are long and flat, and the features perhaps somewhat unpleasing. The foreheads recede, the large,

---

[1] This is very doubtful.
[2] The evidence of some early engravings of Palæolithic cave-dwellers, described in the following chapter, confirms this to some extent. See *L'Anthropologie*, tome vi. No. 2 (1895), Plate V. fig. 4.

bushy red eyebrows meet over the nose, the brows are heavy, and deeply overshadow the eyes beneath. The beards, whiskers, and moustaches vary in style and extent, as such appendages vary now. Many of the women have whiskers, beards, and moustaches. . . . The noses are large and flat, with big nostrils. The teeth project slightly in a muzzle-like fashion; the lower jaws are massive and powerful, and the chins slightly recede . . . . The human creatures are seen to be exchanging ideas by sounds and signs— not by true speech; by chattering, jabbering, shouting, howling, yelling, and by monosyllabic spluttering, sometimes by hilarious shouting (not true laughter), stentorian barking or screaming, or by the production of semimusical cadences. They are also expressing their thoughts by movements of the eyes, eyelids, and mouth, by grimacing, and by gestures made by body, arms, and legs. The men and women have gestures and sounds sufficient for their wants.'[1]

In speaking of the probable mental and moral condition of man in the Older Stone Age, we have, for want of further material, compared him with the aborigines of Australia, Tasmania, and New Zealand, people who are very low down in the scale of civilisation; but, as Mr. Kidd has so well pointed out in his remarkable book 'Social Evolution,' we are very apt to do but scant justice to 'the noble savage,' and to unduly exalt ourselves and our intellectual capacity. We forget that our own civilised state is due to centuries of continuous effort after a higher social state, with greater freedom for the individual and material comfort. And so we give ourselves credit for things in which we have had little or no part, but which are the net result of generations of thought and labour. Our railways, steam-engines, mills, telegraphs, and other appliances of the nineteenth century, are not, in any case, the products of a single brain, but of many, each one taking up the work where his predecessor left off. James

[1] How far this forecast is correct future discoveries must decide.

Watt was not, as some people suppose, the inventor of the steam-engine; he improved on the work of Newcomen and others. Darwin was not the originator of the idea of evolution, for it was present in the minds of ancient Greek philosophers centuries before he was born; and he himself seems to have been attracted to the subject by the speculations of his grandfather, the author of that quaint work 'The Botanic Garden.'

Edison is not the inventor of the electric light; but even when a brilliant discovery is made by a European, we can always perceive that others, by their thoughts and their researches, have paved the way for him. We do not make any allowance for the highly important fact, that whereas we have had the enormous advantage of a long period of stability and continuity, our friend the savage is in a very different position. The thoughts of his ancestors have not been set down in writing; there are not the means of communication which we enjoy; he has no education beyond what he can teach himself as a direct observer of Nature. In fact, while he begins at the beginning, we begin at the end, if such a Hibernicism may be allowed.

An enormous stock of accumulated knowledge and experience comes to us as a kind of natural birthright; and because we can make use of this, we jump to a conclusion very flattering to ourselves, and assume that we are much more highly endowed with intellectual gifts than the so-called lower races with which we frequently come in contact. 'Even these races which are melting away at the mere contact of European civilisation supply evidence which appears to be quite irreconcilable with the prevailing view as to their great intellectual inferiority. The Maories in New Zealand, though they are slowly disappearing before the race of higher social efficiency with which they have come into contact, do not appear to show any *intellectual* incapacity for assimilating European ideas, or for acquiring proficiency and distinction in any branch of European learning. Although they have, within fifty years, dwindled from 80,000 to 40,000, and still continue to make rapid strides on the downward path, the Registrar-General of New Zealand, in a recent report on the condition of the colony, says of them that they possess fine characteristics both mental and physical, and readily adopt the manners and customs of their civilised neighbours. He asserts that in mental qualifications they can hardly be deemed naturally an inferior race, and that the native members of both the Legislative Council and the House of Representatives take a dignified, active and intelligent part in the debates, especially in those having any reference to Maori interests. . . . Even the Australian aborigines seem to provide us with facts strangely at variance with the prevailing theories. The Australian native [1] has been, by common consent of the civilised world, placed intellectually almost at the bottom of the list of the existing races comprising the human family. He has been the zero from which anthropologists and ethnologists have long reckoned our intellectual progress upwards. His mental capacity is universally accepted as being of a very low order. Yet this despised member of the race, possessing usually no words in his native language for numbers above three, whose mental capacity is reckoned degrees lower than that of the Damara whom Mr. Galton compared disparagingly with his dog, exhibits under our eyes powers of mind that should cause us seriously to reflect before committing ourselves hastily to current theories as to the immense mental gulf between him and ourselves. It is somewhat startling, for instance, to read that in the State schools in the Australian colonies it has been observed that aboriginal children learn quite as easily and rapidly as children of European parents; and lately, that 'for three consecutive years the aboriginal school at Remahyack, in Victoria, stood highest of all the State schools of the colony in examination results, obtaining 100 per cent. of marks.' The same facts present themselves in the United

[1] The aborigines of Australia are not yet extinct. Those of Tasmania are practically extinct, being represented only by an old woman, who has been pensioned by the Tasmanian Parliament.

States. The children of the large negro population in that country are on just the same footing as children of the white population in the public elementary schools. Yet the negro children exhibit no intellectual inferiority; they make just the same progress in the subjects taught as do the children of white parents, and the deficiency they exhibit later in life is quite of a different kind.[1] Their real weak point seems to be a want of steady application and persevering energy. The results of steady plodding are truly marvellous, as we see in Germany.

[1] *Social Evolution*, second edition, 1895, p. 293.

## CHAPTER II

### ANCIENT CAVE-DWELLERS

'Necessity is the mother of invention.'

IT is only natural that, in the early stages of human society, caves, caverns, and grottoes should have been regarded with a superstitious awe; and therefore one need not be surprised to find that from this source a long stream of myths and legends has flowed, such as might supply abundant material to the student of folk-lore. In ancient Greece caves were sacred, and oracles were sometimes delivered from them; for example, at Delphi and Corinth. Later on, in France and Germany, we find that they were regarded as the abodes of fairies, dragons, and even devils! Their long passages not unnaturally lent favour to the idea that they led to the regions presided over by Pluto. The River Styx was supposed to flow through a series of caverns leading to the infernal regions.

Many of the stories told in early days of giants and dragons probably originated in the discovery in caves of the limb bones of the mammoth, rhinoceros, and other large animals, associated with heaps of broken fragments, in which latter ignorant peasants saw in fancy the remains of victims devoured by some rapacious monster. During the Middle Ages in Scotland, Ireland, and Germany, the peasants appear to have had a strong belief in dwarfs and 'little folk,' who were supposed to live in caves and underground houses, and from time to time visited the upper regions, especially early in the morning. Many are the stories founded on this belief, and so circumstantially told, that there seems good reason to believe that they are not mere inventions of imagination, but are actually founded on fact. One reads, for instance, of the Celtic people coming to the dwarfs to borrow of them kettles, dishes, and plates, when they were in want of such for some special occasion. There are also stories of conflicts between the two races, and even of the capture of women and children by the smaller race. We shall refer to this interesting subject again in Chapter X., but meanwhile content ourselves with saying, that facts such as these point to a remote time when, in Western and Northern Europe, a small primæval race, driven back by the conquering Kelts, were obliged to take refuge in caves and caverns. Some writers consider them to have been related to the Lapps.

Holy men, such as St. Jerome, wishing to isolate themselves from the world, at times when the dark side of humanity was painfully apparent, so as to make some men despair of the human race, have made their homes in caves, in order to obtain that peace and quiet for which they searched elsewhere in vain.

Whether we regard written or unwritten testimony, we find that caves have played an important part in the history of the human race. Troglodytes, or cave-dwellers, represent the earliest phase with which we are at present acquainted. The first human dwelling was a cave, and probably the earliest tombs were caves. The oldest underground abodes appear to be a kind of reminiscence of the cave-dwelling; and from the rock-hewn tomb of the early Egyptians, with its little chamber where the family assembled from time to time to pray for the departed and offer gifts to the gods, to the shrine with its sacred images, is but a small step. This leads by an easy transition to the rock-cut Temple. Hence might be worked out the whole evolution of sacred architecture, while a corresponding series of stages might be traced representing the ascent of domestic architecture. In all parts of the world caves have afforded shelter and protection to persecuted races and people.

In the history of anthropology, also, caves have played an important part; for, as has been already pointed out, the early researches of Schmerling in Belgium led him to announce the interesting conclusion that an early race of men were contemporary with certain extinct animals, although, at the time, his discovery was not accepted. The subterfuges invented by the learned men of a hundred years ago, in order to avoid this awkward truth, are most remarkable, not to say childish. It was thought in those days that whatever animals lived in the past must have resembled those now inhabiting the world; and the idea of extinct types unknown to man, and unknown to the regions where their bones are found below the soil, was of so novel and startling a character as to appear incredible. Besides, the Mosaic account of the creation made no direct reference to extinct animals, and therefore the notion was not to be entertained. Thus, when the illustrious Cuvier, who founded the science of Palæontology, first announced the discovery of the fossil remains of elephants and other large beasts in the superficial deposits of continental Europe, he was gravely reminded of the elephants introduced into Italy by Pyrrhus in the Roman wars, and afterwards in the triumphal processions or the games at the Colosseum! But he was not to be beaten, and appealed to the fact that similar remains occurred in Great Britain, whither neither Romans nor others could have introduced such animals!

The sagacious Dean Buckland also pointed out that in England, as on the Continent, the remains of elephants are accompanied by the bones of the rhinoceros and hippopotamus, animals which not even Roman armies could have subdued or tamed. Owen also added that the bones of fossil elephants are found in Ireland, where Cæsar's army never set foot. And so that fiction was crushed, but it died hard; and probably there are people living now who would maintain that the mammoth was introduced into Britain by the Phœnicians, while the cave-lion, rhinoceros, hippopotamus, and other creatures, were brought over to take part in gladiatorial shows ! So recently as the year 1882, a writer in a religious journal called 'The Champion of the Faith against Current Infidelity,' in speaking of the discoveries made in the Victoria Cave, Yorkshire, had the rashness to maintain that the bones found therein were those of animals introduced by old Phœnician miners long before the Christian era, and were brought over by these enterprising people in order to use them in working the mines ; but some were used to keep away the Britons ! 'The hippopotamus,' he says, 'although amphibious, is a grand beast for heavy work, such as mining, quarrying, or road-making, and his keeper would take care that he was comfortably lodged in a tank of water during the night !'

Caves have been in all ages used as places of refuge : the reader will readily recall the story of David fleeing from the face of Saul, and hiding in a cave ; of the prophets who, in time of danger, were hidden away in caves ; of British chiefs and nobles who made use of such shelters during the troublous time that succeeded the departure of the Romans from this country, when the Picts and Scots were a terror to the land ; of the celebrated Cluny Macpherson of Cluny, who threw in his fortunes with 'the King over the water,' as Prince Charlie has been called, and in consequence was outlawed, and a price set on his head by the English, so that for seven years he lived concealed in a cave near to his castle in Inverness-shire, and died an exile in France.

But when we pass over the borderland of history and come to read the records of the caves sealed up in cave-earth, stalagmite, breccia, and so on, we find that prehistoric men have for almost countless ages made their abode in caves, and lived therein with all their belongings, instead of in houses, pits, or other structures. This takes us back at once to the Stone Age, which, as already stated, divides itself into two parts—an older or Palæolithic period, and a newer or Neolithic. Our knowledge of the latter period is derived largely from grave-mounds and other remains, as well as from caves, and it will be necessary in this part of our work to leave them out entirely, and confine our attention to the earlier time.

Bone caverns are found in almost all countries where limestone rock abounds, and owe their origin to the dissolving action of rain-water, by a process which can easily be explained. Limestone rock is generally jointed, that is to say, divided by natural lines of division, probably due to contraction after their first formation and drying up. Many of these are more or less vertical, so that rain-water falling on the surface can easily find its way down. The water thus stored up, being ever impelled by gravity, seeks outlets, and forms underground passages and channels by dissolving the limestone as it works along through the joints. This it is enabled to effect by means of the chemical action of the carbonic acid gas which it has taken up from the air in falling to the earth. Thus it comes to pass that in such districts there is quite a system of underground passages, or channels, through which streams find their way once more to the surface of the earth. On descending to these regions the explorer finds a new world of spacious halls, connected together by long, narrow, and often tortuous galleries or tunnels, all made beautiful and fairylike by the presence of countless hanging 'stalactites,' together with curious coloured sheets and curtains of the same material, while on the floor masses of 'stalagmite' rise up to and often meet them. Probably many of our readers are already familiar with the beautiful effects to be seen when such caverns are artificially lighted up. The water which is continually oozing through their vaults falls drop by drop on to the floor, and in this way, by a slight evaporation that takes place, leaves rings of carbonate of lime behind, and these grow and grow, like icicles, until beautiful festoons of hanging stalactites are formed. Cox's Cavern at Cheddar, Somerset, and the Yordas Cavern, Kingsdale, near Ingleton, are famous for their stalactites.

Of late years many geologists have become cave hunters, but the pioneer in that new line of research was the celebrated Dean Buckland, who was the first English geologist to break into this untrodden subterranean world, and see the bones of extinct animals lying undisturbed. In his 'Reliquiæ Diluvianæ;[1] or, Leavings of the Flood,' he gave an account of his labours which fascinated everyone. The title seems a little strange to a modern reader, but it must be borne in mind that in those days people believed that the flood was a universal deluge, and therefore expected to see traces of its effects everywhere. Hence cave deposits, drifts, and river-gravels were all looked upon as records of this great event. But, although this old-fashioned idea has been abandoned, the facts brought to light by this great and genial man remain undisturbed, and his book will always be a classic work. Much has also been accomplished by workers on the Continent, so that to give any complete idea of the state of present knowledge on the subject would require quite a large volume. It will, however, be sufficient for our purpose to say a few words about two or three caverns, such as Kirkdale Cavern, Kent's Cavern, the Bone Cave near Wokey Hole, the Cae Gwyn Cave in North Wales.[2]

[1] *Cave Hunting*, p. 281. The original account is in *Reliquiæ Diluvianæ*. In *Macmillan's Magazine* for September 1862 an account is given by Mr. Taylor.

[2] The reader will find in *Cave Hunting*, by Prof. W. Boyd Dawkins, a complete account of nearly all that has been accomplished of late years in this line of research. The following are the principal British caves which have been explored :—The Victoria Cave, Settle, where the different layers show that it was occupied in Palæolithic times, then in Neolithic, and lastly in post-Roman times. Cox's Cavern at Cheddar (Mendip Hills) is famous for its beautiful stalactites, as also are Yordas Cavern, near Ingleton, and the Clapham Cave in Yorkshire. Besides Wokey Hole and the Kirkdale Cavern the following are noted for containing bones : Pin Hole, Church Hole, and Robin Hood's Cave, of Cresswell Crags, Derbyshire ; The Dream Cave, near Wirksworth, in the same county ; a cave in the Great Orme's Head ; a cavern at Cefn, also at Plas Heaton, near St. Asaph ; one near Tremeirchion, and others at Ffynnon Beuno and Cae Gwyn, in the Vale of Clwyd ; Coygan Cave, near Langharne, Carmarthen, and Hoyle's Mouth, or Oyle Cave, near Tenby ; King Arthur's Cave at Great Doward, Whitchurch, near Ross. The promontory of Gower,

It is strange that, in spite of Mr. MacEnery's declaration that man in Kent's Cavern was contemporary with the extinct mammalia, Buckland could not accept this conclusion. But then everybody believed man to have been 'specially created' about four thousand years ago. The celebrated Dr. Buckland, afterwards Dean of Westminster, was the first among naturalists and geologists to read the records of the Early Stone Age as preserved in the lower cave deposits, and to interpret them to the world at large. It is a fascinating story, full of romantic and weird interest. The pity is that subsequent workers in this department of geology should have so obscured the romance by their 'dry-as-dust' descriptions and ponderous reports of their labours insomuch that no ordinary reader would care to plod through a single chapter of their writings! Not so, however, with the Oxford professor; the account he gives of his labours in the now famous 'Reliquiæ Diluvianæ' reads like a tale of adventure. He was the first to examine these subterranean haunts of primæval men and animals, and so had the advantage of his successors, who, however, lack his facile pen. Be it remembered that, in spite of his mind being warped by the 'Diluvial theory,' he yet was able, by clear cool reasoning, to offer a true solution of the then unsolved problem of the presence, in vast numbers, of fragmentary bones and teeth belonging to creatures many of which are now extinct, while others no longer frequent this country. He proved clearly enough that it was impossible for

in Glamorganshire, contains many caves and fissures with bones. Bosco's Den is one of these; Goat's Hole is another; Longberry Bank Cave, near Penall, in Pembrokeshire. Bones were obtained from fissures in the mountain limestone of Durdham Down, Bristol. And in the Mendips, besides Wokey Hole, are caves at Banwell, Sandford Hill, Bleadon, Hutton, Uphill, &c. A well-known cave at Brixham, Torquay, has been explored by the late Mr. Pengelly and others; also Kent's Cavern. In the same county are caves at Chudleigh and Newton Abbot, at Yealmpton, near Plymouth, and others at Oreston. Fissures containing bones are met with in the oolitic limestone at Bath and Portland.—H. B. Woodward, *Geology of England and Wales.*

the carcase of a hyæna, an elephant, or a rhinoceros to have floated from the region they now inhabit into the recesses of these English caverns. Hence any theory of their transportation by the flood was ruled out of court ; some other explanation had to be found.

On the discovery, in the year 1821, of the Kirkdale Cavern, in Yorkshire, he posted off at once from South Wales to examine it for himself. What did he find there? Not a series of perfect specimens, neatly sealed up in a deposit formed by running water, but a veritable old 'floor' of a cave haunt, where hyænas once lived and feasted ![1] That, at least, was the conclusion he arrived at, and a very bold one it was, considering that few people in those days could look back to anything older than the Flood, much less to a time when the fauna of the country was quite different from what it is now. Worn-out theories of the Flood would not do in this case, for here was a fine red 'loam,' such as is now called cave-earth, and in it were numbers of gnawed and broken bones and teeth. This material was slowly formed by the dissolution of limestone as rain water slowly trickled through from above. It is the insoluble residue of that rock, not a deposit brought by a river from elsewhere. The bones were lying about in little confused heaps ; sometimes they were cemented together by a mass of 'stalagmite.' What a picture the scene presents ! Here were fragments representing some two or three hundred individual hyænas, both young and old. Clearly, then, they came here of their own accord ; and, having come, they intended to stay. Here were the young ones born and suckled, while the old ones were wont to go out on foraging expeditions and bring back their prey. Nor are we left in the dark as to what their prey was. The bison and horse would seem to have been the chief victims to the cunning and greed of these old spotted hyænas.

[1] The bones and teeth were reckoned to have belonged to three hundred individual hyænas.

Doubtless they were often caught unawares as they grazed in the grassy plains of old river-valleys, and we may easily picture the scene. Tall grass in which they fed, and the cunning enemy slowly moving along in packs, getting nearer and nearer to the unsuspecting victim, taking advantage of every bit of shelter, such as a rock, tree, or shrub, until the moment comes for a rush. The horse or bison, as the case may be, kicks or charges, but is bitten and tormented in every direction, until, worn out with fatigue, exhaustion, and loss of blood, he at last lies down to die, and the work of destruction begins. The body is then torn to pieces, and the joints dragged home to be devoured at leisure. But there was no lack of variety in diet, and doubtless the hippopotamus, reindeer, Irish elk, and other graminivorous animals also fell victims, for their remains lie in the red cave-earth at Kirkdale to assure us on this point, and to witness if we lie. But the same evidence tells of other and more powerful flesh-eating animals, such as the mighty cave-bear and the great sabre-toothed tiger.

If anyone should doubt this interpretation of the records at Kirkdale we may cite further evidence, such as cannot be gainsaid, for here are bones actually polished on their upper surfaces by the very tread of the hyænas as they passed around or in and out of their haunt. Their droppings, also, are preserved as 'coprolite,' or *Album Græcum*. But the best part of the story of Buckland's interpretation has yet to come—namely, the way he looked at the broken bones, and, as it were, put them in the witness-box to give evidence. This was of two kinds. First, the marks thereon of hyænas' teeth. Secondly, the fact that they were just those hardest parts of bones that are left by these creatures because they cannot be cracked and eaten. Nor was he satisfied with mere conjecture, which critics might say was only his fancy. To a practical mind like his, nothing short of actual demonstration would suffice. A subpœna must be served on the living hyæna, to come and instruct the court of geologists, and so the shin-bone of an ox was given to one of the hyænas in Mr. Wombwell's Menagerie (it came from the Cape), to see what he would do with it. We will give the account as nearly as possible in Buckland's own words, only simplifying a little here and there.

He was enabled to observe the way in which the animal set to work to eat the bone; it began by biting off with its back teeth fragments from the upper end, swallowing them as fast as they were broken off. On reaching the medullary cavity, the bone split into angular fragments, many of which were caught up greedily and swallowed entire. The creature went on cracking until he had extracted all the marrow and licked out the lowest portion with his tongue. Then came the lower end (condyle), which is very hard, and contains no marrow; it was left untouched. Now this 'residuary fragment' exactly resembles similar bones in the cave! The marks of teeth on it are very few, as the bone usually gave off a few splinters before the creature's teeth had made a hole therein; but what tooth-marks are to be seen entirely resemble the marks left by the old hyænas of Kirkdale Cave. Even the small splinters, in form, size, and manner of fracture, tell the same tale of having passed through the jaws of a hyæna. Buckland preserved all the fragments of this now historic bone for the sake of comparison, and there was no real difference except in point of age. Wombwell's animal left untouched the solid bones of the heel and wrist, and such parts only of the cylindrical bones as are found untouched at Kirkdale. He devoured those very parts which at the cave are conspicuous by their absence. The keeper next morning brought a large quantity of *Album Græcum* disposed in balls, agreeing in shape and size with those in the cave. 'The power of his jaws,' says Buckland, 'exceeded any animal force of the kind I ever saw exerted, and reminded me of nothing so much as of a miner's crushing mill, or the scissors with which they cut off bars of iron and copper in the metal foundries.'

In the Dream Cave, near Wirksworth, in Derbyshire, he came

across a nearly perfect skeleton of the rhinoceros, as well as the bones of the horse, reindeer, and urus, or bison.

The famous hyæna den of Wookey Hole, near Wells, on the Mendip Hills, has been explored by Professor W. Boyd Dawkins, the Rev. J. Williamson, and by others. The main interest of this place lies in the now established fact that it was inhabited by men of the Older Stone Age as well as by hyænas, though, of course, not quite at the same time. Implements of flint, undoubtedly fashioned by the hand of man, as well as two bone arrowheads, were found underneath one of the old 'floors' of the cave in actual contact with teeth of hyænas, and associated with the bones they had fed on, the bones themselves showing every sign of having been deposited there by these animals and not of transportation by water. Here, as at Kirkdale, were to be seen the coprolites, the bones polished on the upper side only by the tread of hyænas ; and, moreover, they would have lost the sharp points they displayed had they been deposited by water instead of being simply left where they now are by the hyænas. Hence all the evidence points clearly in one direction. The number and variety of the bones is remarkable ; teeth are there in even greater profusion (hyænas do not eat them), but the majority belong to the horse.

As at Kirkdale, all the hollow bones were splintered and scored with tooth-marks, and there had been formed a deposit of phosphate of lime on the old 'floor' instead of the balls of *Album Græcum* met with at Kirkdale. Here was evidence of the co-existence with man and the hyænas of such animals as the mammoth, woolly rhinoceros, reindeer, great Irish deer (it should not be called Irish elk, for it is not an elk), cave-bear, lion, wolf, fox. Professor Boyd Dawkins, speaking of his work of exploration here, says of a small passage where bones and blocks of stone were all cemented together in one hard mass : ' The excitement of extracting from these blocks their treasures was of the very keenest, for we could not tell what a stroke of the hammer would

Plate III

HUNTING THE REINDEER IN SOUTHERN FRANCE

reveal. Sometimes an elephant's tooth suddenly came to light, at others a hyæna's jaw or a rhinoceros tooth, or the antler of a reindeer, or the canine tooth of a bear. The bones were so numerous that they scarcely attracted attention.'

The flint implements, calcined bones, and other signs of human occupation were mostly found at or near the entrance, but not in every case. Some of the leg bones of the larger animals, and in particular one of a gigantic urus, have been broken short across, instead of being bitten through   For this Professor Boyd Dawkins offers an ingenious explanation.  He points out that wolves and hyænas at the present day hunt in packs, and often force their prey over a precipice.

'The Wookey ravine,' he says, 'is admirably situated for this mode of hunting, and would not fail to destroy any animal forced into it from the hillside.  It is, therefore, very probable that the hyænas sometimes caught their prey in this manner.  They would not have dared to attack the bears and lions unless they had been disabled.'  At the same time it will naturally occur to the reader that these cunning, cowardly creatures would not have hesitated to attack old animals, or those which might happen to have been wounded in a fight.

The fact that sometimes the bones were found as it were piled up to the roof was somewhat of a puzzle; but the same writer explains this by a reference to floods, which probably took place in these times, whereby water laden with sediment may have elevated the layers of matted bone and other organic material. A succession of floods would in this way gradually fill up any particular passage in the cave.  We cannot do better than quote the Professor's own conclusion of the whole matter: ' All these facts taken together enable us to form a clear idea of the condition of things at the time the hyæna-den was inhabited. The hyænas were the normal occupants of the cave, and thither they brought their prey.  We can realise those animals pursuing elephants and rhinoceroses along the slopes of the Mendips till they scared them into the precipitous ravine, or watching until the strength of a disabled bear or lion ebbed away sufficiently to allow of its being overcome by their cowardly strength. Man appeared from time to time on the scene, a miserable savage armed with bow and spear, unacquainted with metals, but defended from cold by coats of skin. Sometimes he took possession of the den and drove out the hyænas, for it is impossible for both to have lived in the cave at the same time. He kindled his fires at the entrance to cook his food, and to keep away the wild animals; then he went away, and the hyænas came to their old abode. While all this was taking place there were floods from time to time, until eventually the cave was completely blocked up with their deposits. The winter cold at the time must have been very severe to admit of the presence of the reindeer and lemming.'

In the following list[1] the numbers represent jaws and teeth only, and, in the case of man, implements, no human remains having been as yet discovered in the cave.

| | |
|---|---|
| Man . . . . 35 | Woolly rhinoceros . 233 |
| Cave-hyæna . . . 467 | Rhinoceros hemitœchus 2 |
| Cave-lion . . . 15 | Horse . . . . 401 |
| Cave-bear . . . 27 | The great urus . . 16 |
| Grisly bear . . . 11 | Bison . . . . 30 |
| Brown bear . . . 11 | The great Irish deer . 35 |
| Wolf . . . . 7 | Reindeer . . . 30 |
| Fox . . . . 8 | Red deer . . . 3 |
| Mammoth . . . 30 | Lemming . . . 1 |

In Bosco's Den, in South Wales, no less than 750 shed antlers of the reindeer have been found![2] The Bristol Channel was at this time a great fertile plain inhabited by herds of reindeer, horses, bisons, elephants, the rhinoceros, and the hippopotamus,

---

[1] *Cave-hunting*, p. 310.

[2] The number of bones found in some caves is truly enormous. In that of Gailenreuth, Franconia, the remains of 800 cave-bears were obtained. In a Polish cave Römer, some years ago, found the remains of at least 1,000 cave-bears; and from one in Sicily bones of the hippopotamus have been taken weighing twenty tons.

# ANCIENT CAVE-DWELLERS

while the lions, bears, and hyænas haunted the numerous caves. What a different picture is thus conveyed to our minds by the results of geological exploration to that which now meets the eye of the traveller in that part of our country! It has been already stated in the previous chapter that England was at this time united to the rest of Europe, for it is impossible otherwise to conceive how animals from the south could have reached these parts. The North Sea, too, was most likely dry land, and we must try and think of bears or lions walking over what is now the bed of the sea, where our fishermen catch their cod or herrings for our breakfast-table!

In the illustration (Plate I., Frontispiece) (which is based on a photograph, taken for the writer, of the present entrance to the cave) the artist has endeavoured at our suggestion to bring before the reader a scene wherein the man primæval, hard pressed by the unwelcome attentions of a cave-bear, a sabre-toothed tiger, and a hyæna, is defending his hearth and home against these would-be intruders. In one hand he holds the brand which he has hastily snatched from the fire outside; in the other a bone-headed harpoon, which he will presently hurl at one of the animals.

In the background may be discerned the trembling figures of his wife and child, though probably in the Early Stone Age a man possessed more than one wife, and it may even be considered a doubtful point how far she belonged to him exclusively. Instead of giving separate plates to the animals here represented (an expensive way of proceeding), we have requested our artist to put them all into this old 'Eviction scene;' but we are well aware that the three creatures here shown as making a combined attack would not have been on sufficiently good terms with one another to associate together, even for the purpose of turning the man from his home. But, at all events, it makes a more interesting picture, and scientific accuracy may in some cases be pushed too far. Artistic effect has also to be considered. In spite of several supposed discoveries of arrowheads, many authorities whom we have consulted on the subject, both personally and by their writings, doubt very much whether Palæolithic man was acquainted with the bow. This is an interesting point, and one which will be referred to again in our next chapter on the Rock Shelters of Southern France, where some think that certain bone implements, as well as others of flint, may once have been arrowheads. Our friend, Sir Henry Howorth, thinks not. It was, therefore, thought to be wiser *not* to give the man here depicted 'the benefit of the doubt.'[1] Bone needles tell us that his wives could sew skins together for clothing. What his features were like we can hardly tell, having no trace of a skull, at least, from England.

Kent's Cavern, near Torquay, was another famous haunt of Palæolithic man and extinct animals. It has been thoroughly explored by the late Mr. W. Pengelly and others, at the suggestion of the British Association, which gave a grant of money for the purpose. It was at one time occupied by hyænas, lions, bears, and the formidable sabre-toothed tiger (*machairodus*). A certain deposit here known as the 'black band' marks the place where man used to light his fires and cook his food. He seems to have been fairly accomplished for the times in which he lived, for Mr. Pengelly says of him that he 'made bone tools and ornaments, harpoons for spearing fish, eyed needles or bodkins for sewing skins together, awls perhaps to facilitate the passage of the slender needle through the tough thick hides, pins for fastening the skins they wore, and perforated badgers' teeth for necklaces or bracelets. The different layers or strata here met with testify to various changes taking place which are chiefly of interest to the geologist, so we will not attempt to follow them. But below everything else was found what is called a breccia, that is, a layer of angular fragments of rock, with flint implements of a rude type, suggesting that possibly the first occupants of the cave were members of

---

[1] Surely it is time this important point was settled! The writer is inclined to believe they *had* bows and arrows. The reported discovery of flint arrowheads with the human bones at Spy, in Belgium, tends to confirm this view.

# ANCIENT CAVE-DWELLERS

some more ancient race, who were less advanced even in the primitive arts of hunting and making weapons.

No one can examine the list given above of animals living in Britain without perceiving that they present a very motley group, such as in these days would be, to say the least, very unlikely to live in one country and under the same conditions of climate. Here, then, we are presented with a very interesting problem, but at the same time a very perplexing one. Some of the creatures are such as now only inhabit warm southern countries, while others are equally characteristic of cold northern climates. It may well be asked, 'How is this apparent contradiction to be explained?' Have animals in the course of ages altered their habits to such an extent that the hippopotamus and the rhinoceros could live in England amongst the glaciers and snowfalls of the glacial period, while their descendants flourish only in the sunny south, or in equatorial regions? Were they all clothed in fur or wool to protect them from the cold, as has been suggested? This is very unlikely. Again, are we quite sure that the glacial period in Europe was so cold as we have been told by many distinguished geologists? That is a question which will be discussed in a later chapter; and if the climate was, after all, not so very severe, we seem to find in that idea, so lately put forward by certain distinguished geologists, a way out of the difficulty.

Before we go any further it will be well to bring out more clearly the differences between these ancient inhabitants of our country by dividing them into three groups, as given by Professor James Geikie in his well-known work, 'Prehistoric Europe.' We shall see that they comprise a Southern Group, a Northern or Alpine Group, and a Temperate Group, thus:

*The Southern Group:*

| | |
|---|---|
| Hippopotamus | Serval |
| African elephant | Caffer cat |
| Spotted hyæna | Lion |
| Striped hyæna | Leopard |

*Northern and Alpine Group:*

| | |
|---|---|
| Musk sheep | Alpine hare |
| Glutton | Marmot |
| Reindeer | Spermophile |
| Arctic fox | Ibex |
| Lemming | Snowy vole |
| Tailless hare | Chamois |

*Temperate Group:*

| | |
|---|---|
| Urus | Stoat |
| Bison | Weasel |
| Horse | Marten |
| Stag | Wild cat |
| Roe | Fox |
| Saiga antelope | Wolf |
| Beaver | Wild boar |
| Hare | Brown bear |
| Rabbit | Grizzly bear |
| Otter | |

The mammoth, or woolly elephant, may certainly be considered on the whole a northern animal, for thousands of its teeth and tusks are found in Siberia; but, at the same time, it may have ranged pretty far south, and we know it was seen by the reindeer hunters in the Dordogne country when the climate there was glacial (see Plate II.). The woolly rhinoceros must be placed in the same category; but other species, such as *R. hemitœchus*, belonged to a southern group, as did the sabre-toothed tiger. The great Irish deer and the cave-bear are not so easily assigned to their proper place.

A similarly striking difference shows itself when we consider the fauna of the older river-gravels of which we spoke in the first chapter, and it has been suggested that bones found therein may have been washed out of older gravels and redeposited in later ones by the side of bones of creatures that lived later; and in this way northern and southern forms may appear to be associated together when in reality they belong to different times. However, since it appears to be fairly established in the case of the cave deposits that the northern, southern, and temperate groups really were contemporaneous, or nearly so, we see no reason to suppose that the case of the river-gravels is different. At present it is impossible for geologists to determine the exact relationship between the old river-gravels and the cave deposits, but they are usually considered to be in a general sense contemporaneous. Certainly the faunas are the same, and that in itself is a very fair test.[1]

Returning to our cave fauna, the problem it presents can hardly be said to have been satisfactorily solved at present; there are, however, three views which deserve notice. One is that of Professor W. Boyd Dawkins, who was supported by the late Sir Charles Lyell in his belief that this remarkable mixture of forms of life could only be accounted for on the theory that great annual migrations took place, the elephant, the rhinoceros, and hippopotamus working their way up from the South of France during summer as far north as England to feed on the grounds which in winter were occupied by the musk sheep, the reindeer, the arctic fox, and the lemming. If such was the case, some of these African animals must have altered their habits considerably; but that in itself ought not to be a fatal objection, for the rabbits in Australia are now altering their habits, and have taken to climbing trees, and growing longer claws for the purpose! About twenty years ago the bison were still very plentiful in North America. Major Butler describes how enormous herds of them used to go in search of pasture, 'now through the dark gorges of the Rocky Mountains, now trailing into the valleys of the Rio del Norte, now pouring down the wooded slopes of the Saskatchewan. Nothing could stop them on their march; the great river stretches

---

[1] It may be added that in the lowest deposits in some of our caves, such as Kent's Cavern and the Robin Hood Cave, flint implements of very rude make, and resembing those from the higher river-gravels, have been found; while at a less depth implements more skilfully made occur, thus apparently bearing testimony to an advance or progress in the arts. It is believed that the oldest human inhabitants of the cave were contemporaneous, and perhaps identical with, the men of the River-drift.

before them with steep overhanging banks, and beds treacherous with quicksands and shifting bar; huge chasms and earth-rents, the work of subterranean forces, crossed their line of march, but still the countless thousands swept on. Through day and night the earth trembled beneath their tramp, and the air was filled with the deep bellowing of their unnumbered throats. Crowds of wolves and flocks of vultures dogged and hovered along their way, for many a huge beast, half sunk in the quicksand, or bruised and maimed at the foot of some precipice, marked their line of march like the wrecks lying spread behind a routed army.'[1]

Professor W. Boyd Dawkins thinks that the bison was here in the summer, and the reindeer in winter, a conclusion which he thinks is confirmed by a study of the condition of their bones and teeth.

The other theory, put forward by Professor James Geikie, is one which we feel tempted to say was adopted because it fits in with his peculiar theory of warm interglacial periods. He considers that all these animals could never have inhabited our country at the same time, and that when we find their bones associated together, either in caves or in river-gravels, we must suppose that they have got 'mixed up' in the course of time, and that the southern and temperate groups bear witness to warm intervals during the 'Great Ice Age,' as he falsely calls it, when the climate had so far improved as to tempt them to come up from their homes further south, while the northern group marks a stage or stages in that period characterised by extreme severity of climate. We must confess that, to our mind, this view looks very like a piece of special pleading. It is part and parcel of the astronomical theory of the Glacial period, which we shall show in a later chapter to have been recently exploded though the professor still clings to it. The two stand or fall together. If this astronomical theory be true, there *ought* to have been mild intervals in between, like thaws coming in for a few days during a

[1] *Wild North Land*, p. 53.

long frost. Hence his adhesion to an explanation which appears to fit in so conveniently with that theory.

Sir Henry Howorth says there is the clearest evidence that the hyæna fed upon, and was therefore contemporaneous with, the reindeer. He also mentions leaves of the grey willow, which is an arctic species, being mixed with those of the Canary-laurel, and fig (*British Association Report*, Section C, 1889).

Now there is one very obvious way of settling the question which few appear to have pointed out, and it is this. If some explorer of caves could find a bone of a mammoth, or a horn of a reindeer distinctly marked by the teeth of a hyæna (*not* of a wolf or cave-bear), then we should have ample proof that northern and southern animals were actually living together at the same time. The nearest approach we can find to evidence of this kind is a statement by Professor Boyd Dawkins in his work above mentioned.[1] King Arthur's Cave, overlooking the valley of the Wye, in Monmouthshire, is a hyæna den which contains 'the gnawed remains of the lion, Irish elk, mammoth, woolly rhinoceros, and reindeer.' But it is necessary to prove *beyond a doubt* that the tooth-marks are really those of the hyæna (although there is a probability in favour of this). One other piece of evidence ought not to be omitted. It is stated by those who have carefully explored the old hyæna dens of this country that the bones and bony fragments of such different animals as the mammoth, reindeer, cave-bear, urus, horse, bison, red deer, and the lemming, are so associated together on the old floors as to prove that they were all living at the same time. If so, the question would be settled ; and, if the writer may be allowed to express an opinion, he is quite in favour of this view.[2] As will be pointed out in another chapter, the simplest way out of the difficulty is to suppose that

---

[1] *Cave Hunting*, p. 290.
[2] Sir Henry Howorth says there cannot be any doubt about this, and reminds us that he has given many examples in his work *The Mammoth and the Flood*.

the climatic conditions of the time were not unlike those of Switzerland at the present day, or of New Zealand, where glaciers descend into rich fertile valleys where creatures can exist which love warmth, while on the mountain sides, or in the higher valleys and slopes, severer conditions prevailed, and arctic or northern animals and plants flourished. Climate, after all, depends so much on altitude. It used to be considered, not long ago, even by great geologists, that our old cave faunas were *post*-Glacial, and belonged to a time when the glaciers had all melted away. But this view is now largely abandoned, because the evidence appears to be distinctly against it, and it is generally believed that Palæolithic man, and the extinct animals with which he is associated, lived here when all our valleys, as far south as the Thames, were occupied by glaciers, and the higher grounds by snowfields. It has even been suggested that some of the bones discovered in British caves may be of pre-Glacial age; and it is possible that such may be the case. The sabre-toothed tiger (*Machairodus latidens*) is by some thought to be a pre-Glacial and even Pliocene species, but it appears to be quite as closely associated with Pleistocene species as some of the other animals.

The whole question of the true or exact age of the older deposits of British and other caves is still more or less *sub judice*. Dr. H. Hicks, an enthusiastic explorer of caves in Wales, believes that the Cae Gwyn Cave,[1] which he explored and described, affords evidence of having been inhabited by men and extinct animals, either *before* the glaciers came, or before the *supposed* 'glacial submergence;' but this is highly controversial matter, which only highly trained specialists are competent to judge of, and even they are apt to be misled. Professor T. McKenny Hughes, from whom we learned much at Cambridge, takes

---

[1] *Quart. Journ. Geol. Soc.*, 1888, vol. xliv., p. 561. The animals found in this cave are distinctly those of the Ice Age itself (mammoth, reindeer, &c.). Dr. Hicks, Sir H. H. Howorth, and others consider that they have proved this point (*vide* Papers in *Geol. Magazine*).

the opposite view, and considers that the boulder clay seen in this cave is not *in situ*, but has been washed in from similar material that was found above and outside the cave, or, in other words, that it was re-formed, a view many others share. A similar difference of opinion exists with regard to the famous Victoria Cave in Yorkshire, Mr. Tiddeman maintaining that here also there is proof of occupation by extinct animals *before* glacial times, but the evidence is far from convincing. To the outside public probably these differences of opinion with regard to the interpretation of geological 'sections' in caves may appear somewhat strange ; but a section is often the most difficult thing in the world to read aright, and appearances may be so deceptive ! Re-formed boulder clay may look innocent of having been transported away from its original place and deposited again by water running through a cave ! Hence the greatest caution is required.

## CHAPTER III

### THE REINDEER HUNTERS

*The first spiritual want of a barbarous man is decoration.*
CARLYLE, *Sartor Resartus.*

So far, the records of the earliest known men are somewhat scanty. Neither the drift deposits, nor those in British caves, interesting as they certainly are, appear to tell us very much. The reader naturally wishes to obtain a nearer view of these people, to learn more of their manner of life and accomplishments, if they had any. Fortunately, there are other records of great value which give us a much better view of them. These are in the South of France, and take the form of rubbish-heaps, or old 'floors,' in the caves and rock-shelters of the Périgord, first explored by M. Lartet and Mr. Christy.[1]

[1] Their classic work, *Reliquiæ Aquitanicæ*, with full descriptions and plates, was published in 1865-74. Another valuable book, bringing the subject up to date, is now in course of publication, namely, *L'Age du Renne*, by Messrs. Girod and Massenat (Paris). See also *The Deserts of Southern France*, by the Rev. Baring Gould, vol. i., a most interesting work. Also his paper on 'The First Artists in Europe,' in *Good Words*, 1893, p. 100. Boyd Dawkins' *Cave Hunting* and *Early Man in Britain* also contain summaries of the earlier work done in this field. *Der Mensch*, J. Ranke; Lubbock's *Prehistoric Times*; *Exposition Universelle de Paris* (1889); *Histoire du Travail et des Sciences Anthropologiques* (section 1) contain some photographs of groups of models representing reindeer hunters and others, but these are evidently based on the Cro-Magnon skulls, which may be Neolithic. The valuable Journal *L'Anthropologie* (Paris) for 1894 and 1895 contains illustrated accounts of new discoveries from the pen of M. Ed. Piette and others which we have no time now to report, having only recently seen them. Among these are new examples of Palæolithic engravings.

## THE REINDEER HUNTERS

They run along the sides of the valleys of Dordogne and Vézère, at various heights, sometimes far above the level of the river, at other times only a few feet higher.

Here, ages ago, primæval hunters lighted their fires and feasted on reindeer, horse, bison, and other wild animals, most of which are now extinct, throwing away their bones when all their marrow had been extracted, until, in this way, deposits many feet thick were formed. These consist of broken bones of the animals they fed on, reindeers' horns, weapons of flint and bone, layers of charcoal and stones bearing the marks of fire. How many years ago all this took place no one can say, but we believe it could hardly have been less than 10,000 to 15,000 years since the mammoth roamed about these valleys. Now we have unimpeachable evidence that the men of the time saw and hunted that monster, and used his ivory for various purposes. They were so obliging as to make an engraving of him on a portion of a mammoth's tusk!

These old shelters were inhabited in mediæval time, and even now peasants are living there in houses that stand on the top of rubbish-heaps, and casting down their picked bones on to the earthy floor. Caves with similar instruments and bones of Pleistocene mammals have also been discovered in Germany, in Belgium, in Switzerland, and in the Pyrenees. Add to these Great Britain, and it is evident that the wandering hunters of this period inhabited a wide geographical area. It would not be surprising if later researches were to prove that they had the whole of Europe to themselves as far as the wild animals would let them.

Shelters and caverns, or both, are known at Moustier, Solutré, La Laugerie Haut, and La Laugerie Basse, La Magdaleine (Périgord), Bruniquel, the Gorge d'Enver, Les Eyzies, Massat (Ariège), La Vache, near Tarascon (Tarn-et-Garonne), near Tayac (Périgord), Gourdan (Haute-Garonne), and others in France, also Duruthy, in the Pyrenees, and La Salpétrière, in the lower valley of Gardon. (See *L'Anthropologie* for 1894 and 1895.)

The remains naturally divide themselves into bones, weapons, and works of art. Each have their own peculiar testimony to yield. We will take the bones first, because they throw so much light on the conditions of climate under which our primitive hunters lived. Those of the reindeer and horse are very abundant. They split them up and extracted the marrow, which, of course, was highly nourishing. But, as bone was so much used in the making of weapons and utensils, we may suppose that sometimes they were split merely to furnish the necessary material. To these people it was as necessary as wood to a carpenter. Jaws of the reindeer have been found with one or two back teeth struck out. The horns of that useful creature were also utilised, paint-pots being made out of them. At the famous 'station' of Solutré enormous numbers of horses' bones have been found. The animal was not domesticated, but ran about in a wild state in great herds. The men evidently lived here a long time, nor does it appear that they suffered from any shortness of food supplies, for the explorers found literally walls of horses' bones! They were in a broken line 100 metres long, 4 wide, and 3 deep, and it was calculated that they belonged to as many as 40,000 horses! Most were broken; some had evidently been roasted on a fire, and M. Quatrefages mentions that one vertebra has been found pierced by a flint arrowhead.[1]

Professor W. Boyd Dawkins says: 'The cave-men also dared to attack the wild beasts, which were their rivals in the chase. In a sketch in the caves of Dordogne, representing the outlines of a glutton, we have evidence that that animal was familiar to them in Auvergne; and in another, from the cave of Massat (Ariège), that the cave-bear was equally known to the men in the valleys of the Eastern Pyrenees. Vast quantities of broken and split bones in the German caves show that the latter animal formed a large portion of his food in Germany. Among the perforated teeth

---

[1] Some say it is a reindeer's vertebra.

found in the cave of Duruthy are the canines of the great cave-bear. The body must have been cut up, and probably also to a large extent eaten on the spot after the capture of the larger game. For this reason the remains of the mammoth and woolly rhinoceros would naturally be rare in refuse-heaps composed of bones of smaller animals, and to a far less extent of those of the larger, which from their bulk could not be carried. The portions carried off would be cut away from the larger bones. We can picture to ourselves the camp round the carcases, and the fires kindled not merely to cook the flesh, but to keep away beasts of prey attracted by the scent of blood. The tribe assembled round, and the dark trunks of the oaks or Scotch firs lighted up by the blaze, with hyænas lurking in the background, are worthy of the brush of a future Rembrandt. No dogs were used in the chase, and there is no trace of any domestic animal.'

The reindeer also ran wild in great herds; indeed, as far as can be told, no animals were domesticated, not even the dog. The presence of the reindeer is one of the clues by which to judge of the climate at that time; it evidently speaks of considerable cold, especially in winter. In one of the engravings found at La Magdaleine, a hunter, destitute of clothing, is represented as having surprised some horses. Their ears are pricked up as he is about to throw his spear at them. In three cases at least people are represented naked, but for all that, we know that they *did* wear clothing, for they are so represented in another engraving. The horses appear to have had large heads, bushy tails and manes, and probably resembled the ponies of Iceland. The men had great artistic skill, and were very fond of drawing, or rather engraving, on bone or horn, the animals they were wont to chase. Thus we find representations of the reindeer, horse, mammoth, urus, bison, seal, otter, whale (the latter they would hardly have dared to attack), cave-bear, hyæna, antelope, donkey, goat, lizard, and lastly the eel and the pike. The fish were probably speared with small harpoons, perhaps even shot at with

F

the bow and arrow (one wonders whether they had skin boats such as Esquimaux make). Adding together the evidence obtained by the engravings and the actual bones themselves, we obtain a pretty varied fauna, as the following list will show :

Mammoth (*Elephas primigenius*)
Reindeer (*Cervus tarandus*)
Horse (*Equus*)
Aurochs (*Bison europæus*)
Irish elk (*Cervus megaceros*)
Rhinoceros (*R. tichorinus*)
Hyæna (*H. spelæa*)
Cave-bear (*Ursus spelæus*)
Brown bear (*Ursus arctos*)
Cave-lion (*Felis spelæa*)
Sabre-toothed tiger (*Machairodus latidens*)
Ibex (*Capra*)
Saiga antelope
Musk sheep (*Ovibos moschatus*)
Badger
Wolf (*Canis lupus*)
Fox (*Canis vuples*)
Otter (*Lutra*)

Polecat (*Ictonyx*)
Wild cat (*Felis catus*)
Boar (*Sus scrofa*)
Ass (*Equus asinus*)

Snowy owl (*Nyctea*)
Ptarmigan (*Lagopus*)
Willow grouse (an Arctic species)
Wild duck (*Anas*)
Crane (an extinct species)
Capercailzie (*Tetrao*)

Lizard (*Lacerta*)
Eel (*Anguilla*)
Pike (*Esox*)

It will be perceived that the same remarkable mixture of northern and southern types occurs here as in the case of the British cave fauna. And, in spite of theories to the contrary, there can be but little doubt that all, or nearly all, these creatures lived contemporaneously in Southern Europe. The fact that the same artistic race of hunters has left us representations of such different creatures as the mammoth, cave-bear, bison, antelope, and hyæna, seems to prove that, unless some of the works of art were imported from a distance by some simple form of barter, which does not seem very probable, the winters were certainly colder then, but the summers may have been fairly warm. We must picture the valleys invaded by great glaciers, terminating suddenly, as do those of Switzerland, as soon as they reach lower levels where the air is warmer, so that ice melts. There *must* have been a good supply of pasture for the great graminivorous beasts like the rhinoceros or the bison to feed on. Annual migrations probably took place as the seasons changed from summer to winter; but, as suggested in the last chapter, probably some animals kept to the higher grounds, where it was colder, while others, preferring warmth, lived in the plains and lower parts of the valleys.

We turn now to the weapons, utensils, and ornaments of this remarkable people. The vast numbers of flint scrapers found with the *débris*, and also the bone needles, prove that the women spent much of their time in preparing skins for clothing. One needle has been found made out of mammoth ivory. The thread they used was probably made from the sinews of the reindeer, as among the Esquimaux to-day. Visitors to the British Museum should make a point of going to the anthropological department in order to see the fine collection of antiquities of the period we are now considering. There they will see some of the bone needles from the Périgord, with their little round eye-holes very neatly drilled out. A fine flint awl was probably used for this purpose. In the same collection may be seen the spearheads and knives, and saws made of flint, while the examples of work in bone are very numerous. Some of the engravings, or casts of them, are also on view. Many of the needles were made from bones of the feet of the reindeer, being sawn out with delicate flint saws and rubbed down on blocks of sandstone. Some of these have been found with the very grooves in which the needles were ground.

The reindeer hunters daubed their cheeks with red ochre, for at the 'abri,' or station, of La Laugerie their horn paint-pots have been found, as well as little stone mortars, in which they ground the oxides of copper and iron to make the pigment. Nor was a painted face their only idea of ornament; to their credit be it said, pretty sea shells were obtained from the Bay of Biscay or the Mediterranean, and bored so that they could be strung together and worn as necklaces. In one case, at least, fossil shells from

Tertiary strata were used for this purpose. The hearth on which their fires were lighted was a slab of stone.[1]

Pure rock crystals, found in the Auvergne Mountains, were brought to the Vézère. Mr. Baring Gould suggests that possibly they were used for incantation, because crystals are so used by medicine men among the aborigines of Australia.

An engraving of a woman who has been run over by a reindeer shows that bracelets were worn. They seem to have been a hairy people, for in this engraving the woman is evidently intended to be so represented. This is quite what might be expected, and we have reversions to the primitive hairy type in the hairy Ainos of Japan, and a hairy Siamese family described

---

[1] In the year 1863 M. Lartet obtained the assistance of Mr. Christy, a rich English manufacturer, who supplied him with funds to enable him to explore the stations of the Vézère. The result of their joint work for five months was published in *Reliquiæ Aquitanicæ*. Since then others have followed, notably M. Elie Massenat and the Marquis de Vibraye, also M. Filibert Lalande. But much more is yet to be done to unravel these most valuable records of the life of the Palæolithic hunter. Surely, if two or three of the leading societies or academies of England, France, and Germany would combine together for the purpose, sufficient funds might soon be got together to explore some of the shelters which have not yet been touched. Who knows what treasures lie under the soil only waiting for the pickaxe and the spade to bring them to light, and tell us more of the wonderful story of this remote stage of human history? Many are the questions waiting to be solved. We know not with any certainty what were the features and general appearance of this primitive race; where they came from; whether they buried their dead, or whether they had any religion. For all we know there may be engravings under some of the shelters representing their home-life scenes, such as a feast, a burial, a battle, or the erection of a hut or tent. The authors of the above valuable work were pioneers in this field; it is, therefore, not surprising to find that their method was less precise and scientific than that of later workers. It is, of course, most important to note the exact levels at which bones, implements, &c., are discovered; and for this purpose the men who do the digging must be carefully watched. Each layer should be removed separately, otherwise the result is a mere jumble of things, some of which may be Palæolithic, while others may belong to the Neolithic, or the Bronze Ages! As it is, the famous skeletons of Cro-Magnon, which were considered at the time to belong to the old race, are doubtful, and many authorities regard them as Neolithic.

by Darwin. Another engraving of a man stalking a urus also has a row of short straight lines evidently intended to represent hair.

But the masterpiece of art is certainly the wonderful picture of a reindeer feeding (from Thayingen), and now in the museum of Constance. 'Let anyone look,' says Mr. Andrew Lang, 'at the vigour and life of the ancient drawing; the feathering hair on the deer's breast, his head, his horns, the very grasses at his feet, are touched with the graver of a true artist.' Finding the illustrations of this animal in some of the natural history books unsatisfactory, the writer inspected the two specimens in the gardens of the Zoological Society, and his impression was that these early artists have given us a better picture of the reindeer than some of those who now illustrate works on natural history ! It is to be hoped that Mr. Gambier Bolton will turn his camera on to these two specimens to show what they are really like.

In Plate III. (p. 51) the reader will perceive an attempt to render a scene such as may frequently have occurred at this time. It is a reindeer hunt. The men are drawn clothed in skins, with only the arms bare, in hot pursuit of the game, having suddenly surprised a herd on the brow of a gentle slope. They are throwing their bone harpoons. One animal has been hit, the bone end of the harpoon has penetrated his shoulder, and separated, as intended to, from the wooden shaft to which it was attached by a thong of sinew, while the shaft itself is falling to the ground, where it will impede the creature's course, probably leading to its capture.[1] There is a curious engraving of a bison's head and seven human figures, of which three are represented as clothed, but they are so small that nothing else than the bare fact

---

[1] See an interesting paper in *L'Anthropologie*, tom. vi. (1895) No. 3, p. 283, by M. Ed. Piette, ' Etudes d'Ethnographie Préhistorique,' on the use of these harpoons. The author feels bound to point out that the heads of these reindeer are not quite correctly drawn, owing to a mistake on the part of the artist.

can be made out. Possibly, in the excitement of the chase, garments were flung aside (as Highlanders have been known to do in battle), at least, in summer time. Modern Esquimaux, on returning to the interior of their huts, strip themselves entirely free of clothing, in order to allow of perspiration, and they find it quite necessary to do so.

Unfortunately, we have no evidence with regard to the habitations in which these people lived. The rock shelters may have been only temporary camping grounds. Whether they were capable of erecting wooden huts, as has been suggested, or whether they merely made tents of the boughs of trees, covered with skins, it is impossible to say. Perhaps some of the recesses in the shelters were protected against the elements by a kind of screen or curtain made of skins sewn together, as suggested in Plate V.

The entire absence of pottery is remarkable, and it is *believed* that the men of the Older Stone Age were entirely ignorant of the potter's art. The reported discovery of potsheards in association with Palæolithic weapons, &c., must always be looked upon with suspicion. They more probably come from an upper layer of Neolithic Age. Some authorities, however, still believe in Palæolithic pottery (Nadaillac, for example).

The peculiar *bâtons de commandement* made of the antlers of the reindeer, generally perforated with one or more holes, and carved with representations of wild animals, have been the subject of much discussion, and at present it is not known for what purpose they were used. The following explanations have been put forward : (1) That they were the equivalents of the Pogamogan (*puck-à-maugon*) of the River Mackenzie Indians, *i.e.* a *casse-tête* to kill game with. But these implements are *not* pierced with a hole. However, we advise those who may be interested in the question to visit the fine ethnological collection of the British Museum, where they will see something rather similar among the weapons of the Esquimaux which is used to kill captives. (2) Professor Boyd Dawkins thinks they were arrow-straighteners (see his 'Cave Hunting,' p. 355). (3) According to Lartet, Broca, and others, it was a sceptre, or equivalent of the general's *bâton*. The number of holes showed the dignity of the chief to which it belonged. (4) Pigorini considered them to be bridles. This does not seem very likely, and yet among recent discoveries of Palæolithic engravings in France is one representing two horses' heads with rope bridles, figured in '*L'Anthropologie*,' tom. v. No. 2 (1894), paper by Ed. Piette, 'Notes on the History and Classification of Primitive Art.' (5) Reinach thinks the *bâton* might be a trophy of the chase, or the object of some superstitious practices.

Archæologists, following M. de Mortillet, have attempted to trace four distinct epochs in the Older Stone Age, partly by the order of superposition of deposits, partly by the nature of the materials used for making implements, &c., and the degree of skill with which they were made, as follows :[1] —

(1) *The Chellean, or Acheulian*, only partly represented in caves, but chiefly by implements in the old river-gravels. These show very little skill in the making, many of them consisting merely of flint nodules roughly chipped into shape. The spear heads are flat on one side.

(2) *The Mousterian.* These types of implement are proved by their geological position to be clearly later than the preceding, and they show better workmanship.

(3) *The Solutrean*, from the shelter of Solutmé. These show that the flint was removed in flakes by pressure. The edges are sharpened by very careful chipping. The spear-heads are

---

[1] It is hardly necessary to point out that the Older Stone Age represents a long period of time. Some writers consider that the reindeer hunters in Southern France, Belgium, and elsewhere, lived at a period intermediate between Palæolithic and Neolithic times. But, since they saw and hunted the mammoth and many other ancient creatures, we can see no reason for adopting this idea.

quite works of art. Knives, scrapers, &c., also are of improved quality.

(4) *The Magdalenian*, after the cave of La Magdaleine. Flint was then no longer used for spears, arrows, &c.; bone and ivory took its place. This period seems to have marked the highest achievements in art of the reindeer hunters; witness the beautiful engraving of the mammoth.

Sir John Evans, and others, do not consider these divisions to hold good. It seems more likely that the above differences may be otherwise explained; for example, by the abundance of flint in one place and its absence in another, or by differences in skill and artistic taste among different tribes. Among the Esquimaux tribes of to-day, some tribes are better equipped than others.

But, whatever may be thought of the above classification, there can be no doubt at all that the older and the newer Stone Ages are clearly separated from each other. In the first place, the faunas are as different as possible. The Neolithic men were associated with animals that are now living, and they had the dog, goat, pig, &c. Secondly, by the works of art. Neolithic man could make pottery, used the loom for spinning, and was an agriculturist, instead of a mere wandering hunter. And thirdly, by anatomical features. Judging from the skulls from Spy, Neanderthal, Canstadt, Naulette, and Trou du Frontal, the skull of Palæolithic man belongs to a ruder and altogether more primitive type. But there is not very much to be said on this subject at present, because of the scarcity of skulls that can be shown to be *undoubtedly* Palæolithic. And fourthly, by the different levels at which relics of the two periods occur. It is always in the upper layers of cave deposits that implements of Neolithic man are found, often overlying those of the earlier period. In the case of river-gravels, Palæolithic implements are found in the higher gravels, while Neolithic ones occur in the newer gravels down below, the rivers having since cut down their beds to lower levels.

Thus it will be seen that there is no direct passage of the one set of deposits into the other, but a great gap always occurs, signifying a long interval of time in which great geographical and other changes took place. It is said that in some caves in the Pyrenees layers of one age pass without any break into those of the other. But this is by some considered doubtful.

M. Bergounoux, schoolmaster at St. Gevy, discovered some stations in a magnificent precipice near the railway station of Conduche. Speaking of another shelter in the Cele, he says:—
'A careful scrutiny of the locality, and of the objects exhumed, allows us to picture the aspect of this ancient human station. Under the screen of rock are to be found quantities of barbed arrow-heads, needles, spear-points, &c., of bone. Outside the shelter these people worked at the flints. Many blocks of stone seem to have served as seats. Some of the men squatted cross-legged on the soil, working diligently, their minds tranquillised by the beautiful landscape before them. Hither and thither ran the women and children, making a considerable noise. The former had not only the care of the infants, but also the preparation of the meals. The hearths are very numerous, and the ashes form a bed that extends for some distance from the base of the cliffs almost without intermission. Probably the fires were kept in by night as well as by day, to ward off the wild beasts. The shelter, that is to say, that portion sheltered by the rock, was probably extended artificially by means of tents of skin, even by rudimentary huts (see Plate IV.). Possibly, numerous cabins, back to back, covered the space now black with hearths, and, in that case, this represents a congeries of fires forming a sort of village.' The following picture is from the pen of Professor W. Boyd Dawkins:

'The folk of that closing period lived very much in the same way as the Eskimo live now, fishing in the cold waters and hunting in the "barren grounds;" the refuse of their feasts was allowed to accumulate on the floor of their dwelling places, and they probably suffered no more inconvenience from the presence of the unsavoury heaps than similarly circumstanced tribes in our own day. We can picture them to ourselves feasting round their fires on reindeer flesh, or splitting up the bones and sucking the juicy marrow. At other times, when perhaps reindeer hunting had proved unsuccessful, they were content to catch such fish as they could in the rivers, or to capture lemmings, weasels, water-rats, and other small animals and birds. Their tastes do not seem to have been very eclectic, and from the relics of their feasts we gather a pretty fair idea of the mammalian fauna of the lands they lived in. But, as we have seen, they seem to have had no domestic animals, nor have we any reason to believe that they knew anything of agriculture. The potter's art appears likewise to have been unknown. The most distinguishing characteristic of the reindeer hunters, however, was their love of art, a characteristic which, as we know from the analogy of the living Eskimo, may co-exist with a very low state of civilisation.'

How far scientific men are justified in drawing conclusions with regard to the appearance and physique of a whole race of prehistoric men from two or three skulls, much less from only one, is a doubtful question, but it may be stated that M. Quatrefages, from a study of the Cro-Magnon remains (which he takes as typical of the race), concludes that the reindeer hunters of the South of France possessed 'a fine open forehead, a long, narrow and aquiline nose, which must have compensated for any strangeness which the face may have acquired from the smallness of the eyes; and these men, who were far from being ugly, combined a high stature with powerful muscles and an athletic constitution, and were therefore in every way fitted for struggling against such perils and difficulties as they most undoubtedly had to encounter.' This conclusion is not generally accepted now, on account of the uncertainty with regard to the exact position, or level, at which the bones and skulls were found. Cro-Magnon is on the banks of the Vézère. The remains belong to a man, a woman, and an unborn child. Professor Boyd Dawkins, who has visited the cave, says distinctly that they are interments of a later age, and occupy a higher level than the hearthstones of the reindeer hunters.

Every work dealing with prehistoric man contains long and tedious descriptions of some of the famous skulls which have been supposed to belong to the Older Stone Age. We will, therefore, avoid these technical details, and merely point out that, with the exception of the skulls found at Spy—and perhaps the Neanderthal and Naulette specimens—there are none which can with any certainty be assigned to a period more ancient than the Later Stone Age. Among the category of doubtful cases must be included the skulls, or skeletons, from Cro-Magnon, Candstadt, Neanderthal, Naulette (a lower jaw only), Trou du Frontal, Engis, Gendron, Gailenruth, Aurigac, Solutre, Balzi Rossi (near Mentone),[1] the 'crushed man of La Laugerie' (which M. Massenat insists is a genuine case), Duruthy (in the Pyrenees), Bruniquel (in Glamorganshire), and the Paviland Cave.

[1] After studying the detailed accounts given in the *Anthropological Journal*, vol. xxii. p. 287, and *Natural Science*, vol. i. (1892) p. 272, we find ourselves unable to accept these as examples of Palæolithic interments. *L'Anthropologie*, tom. vi. No. 3 (1895), p. 314, contains a review (with illustrations) of a work by two French authors, who claim to have discovered a Palæolithic interment. But, judging from this review, we are bound to say that they do not give good reasons for their belief. The work is entitled *Les hommes préhistorique dans l'Ain*, broc. 8, avec 7 pl. (Bourg, 1895). It seems more probable that burials did *not* take place among the reindeer hunters. But the Neolithic men who succeeded them *did* undoubtedly use the same places for their interments. Hence, unless it can be clearly shown that there is no disturbance of the layers as originally formed, it is useless to argue from the discovery of a skeleton associated with the bones of reindeer, &c., that it is therefore of Palæolithic age. The greatest care is required in exploring these old sites. *Natural Science*, vol. vi. p. 368 (June 1895) reports the discovery by a well-known archæologist, Mr. K. J. Maska, of ten almost complete skeletons buried in brick earth (loess) of Predmost, in Moravia, with remains of mammoth and arctic fox in their immediate neighbourhood. But as they are interred under a slab of stone, the case seems somewhat suspicious.

The division of prehistoric time into three periods characterised by the use of stone (or bone), of bronze, and of iron, though it has its advantages, is certainly not without drawbacks. In any given district, when deposits of all the three ages are found directly overlying one another, it may be very useful. But some archæologists have been misled by adhering too strictly to the scheme, and it should always be remembered that, when one district is compared with another, especially when they are widely separated geographically, great differences in the habits and culture of the people must be expected.

To take an illustration from the present day, it might have been said a short time ago that some of the people in Iceland had not advanced beyond the culture of the Stone Age; for in certain of the more remote parts of the island articles both of stone and of bone were very largely in use. Dr. Tempest Anderson saw a wheelbarrow with a stone wheel, a steelyard with a stone weight, a stone hammer, and a net with bone sinkers. At the same time a quern (or flat stone) was in use for grinding corn, also horn stirrups, harness fittings of bone, and bone needles. With regard to the latter, it may be mentioned that they were in use down to the reign of 'Good Queen Bess!'

The North American Indians, at the time of the Spanish invasion, were practically living in the Stone Age. Some tribes had discovered native copper, and flattened it to make weapons with. They did not pass through a Bronze Age, and, even now, though they can get iron by barter, it is comparatively rare, stone axes and arrowheads being in common use (as may be seen from a visit to the fine Ethnological Gallery at the British Museum).

With regard to the way in which a flint knife is made, we cannot do better than give Mr. Baines' description of the way in which the natives of North Australia proceed, without using a stone hammer. He says: 'The native, having chosen a pebble of agate, flint, or other suitable stone, perhaps as large as an ostrich egg, sits down before a larger block, on which he strikes it so as to detach from the end a piece having a flattened base for his subsequent operations. Then, holding the pebble with its base downwards, he again strikes so as to split off a piece as thin and broad as possible, tapering upward in an oval or leaf-like form, and sharp and thin at the edges. His next object is to strike off another piece nearly similar, so close as to leave a projecting angle on the stone, as sharp, straight, and perpendicular as possible. Then, again, taking the pebble carefully in his hand, he aims the decisive blow, which, if he is successful, splits off another piece with the angle running straight up its centre as a midrib, and the two edges sharp, clear, and equal, spreading slightly from the base, and again narrowing till they meet the midrib in a keen and taper point. If he has done this well, he possesses a perfect weapon, but at least three chips must have been formed in making it, and it seemed highly probable, from the number of imperfect heads that lay about, that the failures far outnumbered the successful results.'

Rude as some of the Palæolithic flint weapons may appear, yet no doubt they were useful for a variety of purposes, such as cutting off branches of trees, grubbing up roots, killing and cutting up animals caught in the chase, scraping skins and making garments out of them, chopping up firewood, carving wood or bone handles for knives and daggers, making shafts for spears. Some of the large roughly chipped implements from the Thames Valley (in the British Museum) are not easily explained; they seemed more suitable for making holes in the ground than for anything else.

Judging from the habits of modern savages, we may conclude that the men of this age were far from neglecting other materials; probably they found plenty of use for ivory, bone, wood, shells, horn, and hide. The first spoon was probably a sea-shell attached to a cane or wooden handle. The Negrittos of the Andaman Islands, one of the most primitive races known, use bamboo for nearly everything.

One would like to know how fire was obtained. The men of the rock shelters appear to have used lumps of iron pyrites, which they struck with flint. Some savages rub two sticks together, others use the fire-drill. Speaking of fire, Professor Fraser thinks that the keeping up of perpetual fires, and the ministrations of vestal virgins, had their origin in those primitive times when the fire was in the chief's tent or hut, and his wives and daughters tended it, and so in the course of time they became priestesses of the sacred element. Among the Damaras of South Africa a perpetual fire is kept burning about the chief's hut. Whenever the headman of a clan, or kraal, was about to move away to some distance a portion of this holy fire was given to him with which to set up a hearth in the new village. This leads to the subject of cooking, and the question naturally arises how it was invented, the answer to which can only be guessed at. Probably it was in some accidental manner. A man or woman may have left a piece of meat on a hot stone near the fire, or on a hearthstone, after the fire had gone out. Curiosity, and the attractive smell of a grilled steak, may have led him to taste, and, having tasted, he would be sure to try the experiment again. Next time he would try and roast the joints over the fire, suspending them on sticks. Having made this great discovery, he would naturally communicate the result to his chief, who doubtless would call the whole tribe together to a grand feast to celebrate the auspicious occasion, truly an epoch in human progress! Music and dancing, we might well imagine, would be invoked to add to the hilarity of the proceedings. But, whatever the way in which the culinary art arose, the presiding genius of the hearth, or table, was always a woman.

> Whose altar was the cheerful table spread,
> Whose sacrifice the pleasant daily bread,
> Offered with incense of sweet childhood's mirth
> And parent's priestly ministrations worth
> More than all other rights that ever shed
> Light on the path that those young feet must tread.
>                                    MARGARET G. PRESTON.

The more primitive the condition of man, the less power he possesses of repressing those evil passions of envy, malice, and hatred which lead to murder and warfare. It is, therefore, not to be expected that the Stone Age was a kind of prehistoric millennium, rather should we expect some day to find that fighting took place between different tribes. There may be battle-scenes engraved on bone or horn among the layers of *débris* in the rock shelters of Périgord or elsewhere. The women were probably brave and courageous.

The customs of modern savages throw some light on these matters. Travellers say that in Samoa the wives of chiefs and principal men generally follow their husbands wherever they might be encamped, to be ready to nurse them if sick or wounded, true precursors of the hospital nurses who now invariably accompany our military expeditions and render such valuable service. A heroine would even follow close upon the heels of her husband in actual conflict, carrying his club and some parts of his armour. Tacitus said of the Germani: 'It is a principal incentive to their courage that these squadrons and battalions are not formed by men fortuitously collected, but by the assemblage of families and clans. Their pledges are also near at hand; they have in hearing the yells of their women and the cries of their children. These, too, are the most revered witnesses of each man's conduct, these his most liberal applauders. To their mothers and their wives they bring their wounds for relief, nor do they dread to count or to search out the gashes. The women also administer food and encouragement to those who are fighting.'[1] He also says that armies beginning to waver have been rallied by the females, through the earnestness of their supplications.

Here is another picture to help us to realise the condition of men at this stage in human progress.

---

[1] Tacitus, *Germania*, vii. and viii.

Speaking of the cave-dwellers by the Red Sea, Diodorus Siculus says : ' Mocking at all manner of sepultures, for as soon as any of them is dead, they tie his head between his legs with a withe of hawthorn or willow, and dragging the corpse to the highest place they can find, with laughter and jeering they overwhelm it with stones, and then, putting a goat's horn on the top of the stones, they leave it there without any pity or compassion at all.' The following picture from the pen of Tacitus of the Finnish tribes in his time seems admirably adapted to the ways of the people of the Older Stone Age. 'They are wonderfully savage and miserably poor. Neither arms nor homes have they; their clothing is skins, their bed the earth. Their arrows, for want of iron, are tipped with bone. The women live by hunting, just like the men, for they accompany the men in their wanderings, and demand a share of the prey; and they have no other refuge for their little children against wild beasts or storms than to cover them up in a nest of interlacing boughs. Such are the homes of the young, such the resting place of the old. Yet they count this greater happiness than groaning over field labour, toiling over building, and poising the fortunes of themselves and others between hope and fear. Heedless of men, heedless of gods, they have attained that hardest of results, the not having so much as a wish.'

And lastly, the question naturally presents itself, 'Had these early hunters any religion?' At present, we regret to say, there is hardly a particle of evidence of any kind to enable us to form a conclusion on the subject. We are far, however, from saying that the question will never be answered, for it is quite within the limits of possibility that future researches will bring to light facts bearing on the subject. On *à priori* grounds, one would expect that a people in whom the artistic faculty was so well developed would at least have some ideas, however crude, on the subject of the unseen world ; but the absence of evidence forbids speculation. Mr. Benjamin Kidd and others have done good service in pointing out the extreme difficulty of obtaining from travellers and missionaries trustworthy evidence with regard to the religious ideas of the most primitive races.

In the first place, it is necessary to start with a definite idea of what we mean by religion, for many of the conflicting statements met with in works of travel are due to a want of clear ideas on this subject. Another serious obstacle which lies in the path of most travellers is the fact that the answers given to their inquiries are often given through an interpreter, who, of course, may easily fail to give a proper interpretation. And in some cases the traveller trusts entirely to his own knowledge of the language, which probably is far from sufficient for the purpose. Nor are these the only difficulties, for primitive peoples, when suddenly confronted with questions put somewhat abruptly by a European traveller, are apt to give misleading answers, and perhaps even to purposely deceive the inquirer.

To show how great are the divergences of opinion on this subject we cannot do better than quote, as Mr. Kidd does, some remarks of Max Müller's. 'Some missionaries,' he says, 'find no trace of religion where anthropologists see the place swarming with ghosts, and totems, and fetishes; while other missionaries discover deep religious feelings in savages whom anthropologists declare perfectly incapable of anything beyond the most primitive sensuous perception.' And again he says: 'When the missionary wants to prove that no human being can be without some spark of religion, he sees religion everywhere, even in what is called totemism and fetishism; while if he wants to show how necessary it is to teach and convert these irreligious races, he cannot paint their abject state in too strong colours, and he is apt to treat even their belief in an invisible and nameless god as mere hallucination. Nor is the anthropologist free from such temptations. If he wants to prove that, like the child, every race of men was at one time atheistic, then neither totems, nor fetishes, nor even prayers or sacrifices are any proof in his eyes of an ineradicable religious instinct.' The different observers do not seem to know what to look for as the essential element in religion.

Now the life of a savage is, it seems, carefully regulated by curious customs which, however unnecessary or foolish they may seem to the 'enlightened' product of civilisation, are yet of vital importance to him. Of such customs it cannot be said (as is too often the case with our own) that they are 'more honoured in the breach than in the observance;' for they are as precise and definite as, say, the 'Rules and Regulations' for the street traffic of London! And, like the latter, they are the means by which chaos is avoided and things made to work more or less smoothly. These unwritten laws, then, are evidently the prototypes and forerunners of those highly complicated legal and religious codes which are characteristic of all civilised communities. Fear of the supernatural is the potent force by means of which the customs of the savage are enforced. He surrounds himself with many ghosts, spirits of ancestors, of the woods, trees, air, earth, sky, and water, and these must be propitiated and obeyed. The priest or magician who interprets their wishes plays, as it were, the part of the policeman in a civilised community: both must be obeyed, or punishment follows swift and sure. The 'supernatural sanction,' as Mr. Kidd well shows, is the essential feature of all religions.

Now, if there are such difficulties in the way of ascertaining the religious ideas of living men, how much greater must be that of the anthropologist when he sets to himself the task of endeavouring to find out the ideas of dead men on the subject, especially when, as in the present case, they lived so long ago!

May we be permitted to recall the words of Pope, and, applying them to the Palæolithic hunters of the Dordogne, express a hope that even they were not without some hope of a life beyond?

> Lo, the poor Indian! whose untutored mind
> Sees God in clouds, or hears Him in the wind;

HUNTERS FEASTING ON HORSE-FLESH IN SOUTHERN FRANCE

His soul proud Science never taught to stray
Far as the solar walk, or milky way ;
Yet simple Nature to his hope has given,
Behind the cloud-topt hill an humbler heaven,
Some safer world in depth of woods embraced,
Some happier island in the watery waste.

For reasons already stated, the natives of Tasmania, as observed by travellers in the earlier part of the century, would appear to be modern representatives of the men of the Older Stone Age. They have been described for this purpose by Professor Tylor.[1] Their religion was a rude 'Animism ;' they thought a man's shadow was his ghost. The echo of his voice, when he spoke against a cliff of rock, was his shadow talking ! They wore the bones of dead friends in order to secure the protection of their spirits. They believed in a future state of which the abode was some distant region of earth. Foreigners were identified with dead Tasmanians returned from the spirit land ! Primitive as all this may seem, it does not take us back to the first beginnings of language, religion, or society.

[1] *Anthropological Journal*, vol. xxiii. p. 151.

---

NOTE.—Just before going to press we hear that M. Rivière has discovered a cave in Dordogne with numerous engravings of animals deeply cut on the walls. Some of them are covered by a layer of stalagmite.

## CHAPTER IV

### THE MYTH OF THE GREAT ICE SHEET, AND THEORIES OF THE FLOOD

> A huge stone is sometimes seen to lie
> Couched on the bald top of an eminence:
> Wonder to all who do the same espy,
> By what means it could thither come and whence,
> So that it seems a thing endued with sense,
> Like a sea beast crawled forth, that on a shelf
> Of rock or sand reposeth, there to sun itself.
> 
> WORDSWORTH.

SCIENCE and superstition have, as a rule, not much in common, they are generally deadly enemies; and it is a frequent saying that the light of science scatters the dark mists of error and superstition. And yet they have this in common, that time-honoured fancies, which have come down to us from the dark ages, can often be shown to take their origin from established facts. As we have already seen, it is an established fact that caves frequently contain the bones of monsters that lived long ago. In the absence of scientific explanation, superstition stepped in to account after its own fashion for the phenomenon. Hence the stories of dragons living in caves, where they devoured their victims. In the present chapter we shall deal with another set of facts which, in the world's infancy, were explained by giants, devils, fairies, elves, witches, *et hoc genus omne*.

Over a large part of Europe, thousands of great boulders lie on the ground, evidently wanderers far away from their original homes. The people of old, having a love of the wonderful and supernatural, in order to explain this, did not hesitate to call in

## THE MYTH OF THE GREAT ICE SHEET 85

the agency of supernatural beings, who were supposed, in various ways, to have left them there. Thus, the numerous 'erratics,' or wandering boulders, of the Scottish Lowlands have given rise among the peasantry to all sorts of strange fancies, some of which doubtless are survivals from the Stone Age. Sir Archibald Geikie, in his delightful book on 'The Scenery of Scotland,' alludes to them as follows : 'Many a wild legend and grotesque tale of goblins, witches, and elfins has had its source among the grey boulders of a bare moor. "Giants' stones," "Auld Wives," "Lift," "Witches' Stepping Stanes," "Warlocks," "Burdens," "Hell Stanes," and similar epithets, are common all over the Lowland counties, and mark where, to the people of an older time, the singularity of these blocks proved them to be the handiwork, not of any mere natural agent, but of the active and sometimes malevolent spirits of another world. Nor need this popular belief be in any measure a matter of surprise ; for truly, even to a geological eye, which has been looking at the same phenomena for years, each fresh repetition of it hardly diminishes the interest, nay, almost the wonder, with which it is beheld. We have disposed of the old warlocks and brownies ; and yet, though we can now trace, it may be, the source from which the stones were brought, and the manner in which they were borne to their present sites, their history still reads like a fairy tale. There they lie crusted over with mosses and lichens ; tufts of heather and harebell and fern nestle in their rifts, while all around perhaps is bare, bleak moorland. How came they there?'

It is rather late in the day to say that the agency by which these strange wanderers, or 'erratics,' are now accounted for is ice. Everybody knows that, or ought to, for it is taught in our schools, and we find it in popular guide books, even in our newspapers. The valleys of Scotland, the North of England, and Wales were once full of ice, and glaciers crept slowly along, bearing these stones with them ; till a time came when the climate changed, the cold diminished, glaciers slowly retreated, leaving their stony burdens in the valleys and on the lower slopes and hillocks.

But, strange to say, some of the leading geologists of the century have been so led astray by the many and striking signs of the action of ice all over Northern Europe and America, that, in their fervid imagination, glaciers have grown to a universal ice-sheet. The idea of what they are pleased to call a 'Great Ice Age' has possessed them like an evil demon. Hence came a deal of vain talk about a great Polar ice-cap creeping down over Europe from the Arctic regions! Sober minded men of science were astounded, and could only say to themselves, 'This way madness lies.' To invoke such an unheard-of agency seemed like an appeal to some weird supernatural power. They refused to accept this new bugbear as the peasants of old believed the legends that are now fast dying out.

The ordinary reader may well inquire whether Science is so entirely free from wild fancies and marvellous theories as to be able to boast that she, at least, has no superstitions! In the face of certain extravagant hypotheses which have been brought forward within the last fifty years or so, it must be confessed that such is not the case. In the particular instance under consideration an extravagant theory has become, with certain persons, little less than a superstition; at least we may, with Sir Henry Howorth, call it a nightmare. Hence it must be sorrowfully admitted that, in some cases at least, Science has not only failed to scatter the mists of ignorant superstition, but has gone out of her way to create misty speculations which to most of us cannot but appear as superstition. Forty or fifty years ago this great myth was created.

The idea 'caught on,' and a whole school of ardent glacialists—Agassiz, Ramsay, James Geikie,[1] Croll, and others went mad over ice. This was to be the great undiscovered agent that could do

[1] Prof. J. Geikie has now abandoned this theory; see his address, Geol. Section Brit. Assoc. 1889, reported in *Nature*, vol. xl. p. 486.

everything ! Such feats as carving out valleys, scooping out hollows for future lakes of great depth, transporting countless boulders of all sizes for hundreds of miles east and west from Scandinavia, to England on the one side, and Russia or Poland on the other, they considered as mere trifles, only some of its playful little ways ! In some unaccountable manner ice became a kind of select scientific cult—a new religion, in fact. The glacier and the ice-sheet were objects of worship, endowed with mysterious and awe-inspiring powers, before which geologists must prostrate not their bodies but their minds. At the feet of this new idol men of great ability and learning laid down their offerings with a childlike simplicity that was touching were it not so ludicrous ! To reason with them was hopeless ; they had apparently given up the power of using that faculty, and relied on a fervid imagination, which had usurped its place. Now, conclusions not based on reason must fall to the ground and be shattered ; 'all the king's horses and all the king's men cannot set them up again.'

The reader will, however, be glad to learn that there are many sober-minded geologists who have not bowed the knee to this Baal, or invoked his power to deliver them from a dilemma. Cautious scientists like Professor Bonney,[1] and Matthieu Williams in England, and Pettersen and Kjerulf in Norway, have for years been strenuously opposing the theory, and have succeeded in putting reasonable limits to the power of ice and the former extent of the old ice-sheet. These, together with that valiant but most courteous fighter, Sir Henry H. Howorth, have to some extent been successful in their endeavours to drive this terrible monster from the field of geological speculation, and the myth of the great ice-sheet may be said happily to be dying, as all myths eventually do, though apparently it dies hard.[2]

[1] See Prof. Bonney's latest work on *Ice-work Past and Present* (International Scientific Series).
[2] The student may consult numerous papers of the last ten years in

'The fact is,' he says, 'we have been pursuing a "will o' the wisp," instead of a real induction in following the lead of the cultivators of the great Glacial myth. We must countermarch, that is plain; we must get rid altogether of the notion that half the Northern temperate zone was swathed in ice and snow, and realise it as it may still be realised in New Zealand, in the Himalayas, and the Altai Mountains, where glacier and forest are almost conterminous. This view entirely does away with those theories according to which England, &c., &c., was stripped of their verdure and their living inhabitants, and encased in a suit of frozen armour, and were re-invaded and re-peopled by fresh importations of plants and animals which had meanwhile retired goodness knows where, and returned goodness knows whence, without having varied, in a mode at once simple and astounding' ('Geological Magazine,' 1893, p. 365).

But erratics are not the only products of the action of ice. We must now say a few words about a remarkable deposit known as till or boulder clay, which is to be met with in the places where glaciers have once been. It may be described as a deposit of very tough dense clay, with a great many stones in it of all sizes; and these, be it noted, are not in any way arranged, but big and little ones occur together, and at all angles, *suggesting that they have all been forcibly squeezed together*,[1] rather than deposited by water. It is a simple matter to calculate the pressure resulting from a given depth of ice, and it has thus been found that at the bottom of a glacier 1,000 feet thick the pressure must be about twenty-five tons to the square foot, or nearly 400 lbs. to the square inch. Cutting through till is very hard work, as has often been found in making railway cuttings. The stones are

the *Geological Magazine*, by Sir H. Howorth, on this question; also his big book, *The Glacial Nightmare and the Flood*, which contains a most exhaustive history of the growth of this idea, and an able criticism of it.

[1] In some places in Scotland till presents the appearance of having been forcibly squeezed into strata over which it has been dragged by the ice, so that the underlying strata contain veins of it.

not exactly rounded (as they would be by water action), but have their angles rubbed off, and their flatter surfaces rubbed down and scratched, or 'striated,' as those in modern moraines are often found to be.

These and other facts seem to show clearly enough that till was in *some places* formed under ice. The stones were embedded in the lower parts of the glacier, and, coming into contact with bare rock underneath, were slowly rubbed down, all the while being firmly embedded in the ice. What thus happens to the rocky bed of the glacier is even more striking, for projecting rocks are smoothed and rounded down till they become the well-known *roches moutonnées* of geologists (p. 92). Another feature of this peculiar deposit is that it always partakes of the local character of the rocks in the district where it is found. Thus in one place it will be dark, clayey, and very stiff, the rock in that district being dark clay ; in another it will be reddish and light, because there the prevailing rock is a red sandstone. What better proof, then, could one want that this perplexing deposit, about which so much difference of opinion has been expressed, has been often formed or rather accumulated under glaciers ? Much of the material was doubtless ready to hand at the very beginning of the Glacial period, in the form of pre-Glacial (Pliocene) soil, and accumulation of rocks (screes, &c.), such as are always forming among mountains. So it got embedded in the ice. Most geologists point to the till as the actual product of the grinding action of an ice-sheet. To a certain extent abrading action does go on under glaciers ; that is proved by *roches moutonnées*. But it is greatly exaggerated, and the other obvious source of clay and stone is generally neglected.

No deposit laid down under water would show such a curious jumble of stones, &c., nor such toughness as some boulder clays do. It evidently has not travelled very far. When the ice finally melted, these deposits, of which the materials were embedded in the glaciers, were left behind. They are, however, not the same as moraines. Although till is unstratified, that is, not arranged in layers like ordinary sedimentary deposits, yet it often has stratified layers of sand and gravel intercalated with it. Also layers of peat, bones of the mammoth, reindeer, and other mammals have been dug out of it, not to mention trunks of trees. The best explanation of such facts appears to be that the old glaciers of the period now and then encroached over lakes and watercourses, burying up their layers of peat gravel or sand with their own peculiar deposit, the boulder clay or till. Sometimes there was an advance of the old ice-sheet; sometimes it receded, not, as some writers maintain, on account of any particular change of climate (inter-Glacial period), but because the snowfall had increased or diminished. For it is snow falling on the mountains that forms the *névé* from which a glacier flows.

But the conditions were not always the same. Evidence collected in England and Scotland proves that, in some places, the old glaciers reached the sea, and even pushed their way into it, for marine shells have been found in boulder clay in Lancashire and elsewhere. Some writers have taken this to mean that the land was submerged, and bring it forward as an argument for their theory that boulder clay must have been formed by icebergs depositing their burdens of mud and sand in the sea. This view, once so generally accepted, has been partly abandoned, as will be seen further on. To sum up, the toughness of till, its peculiar jumble of stones of all sizes, the local character of the material composing it, the absence of stratification, and, lastly, the phenomenon known as 'crag and tail,' all go against the old theory of icebergs or of coast ice. But if we go back to the early history of this controversy about the drift deposits, we find that all the older observers, naturalists, and travellers were great believers in water. That was the element then in vogue, the fashionable fetish. Gradually it was perceived that the facts could not be thus explained, and geologists, following Agassiz and others, became quite as enthusiastic believers in ice. Of course they were quite right in attributing many of the drift phenomena to this agent, our old friend water in a new shape. But the mistake they made was to grossly exaggerate its power. They jumped, not 'out of the frying-pan into the fire,' but from a sea of rushing waters on to 'thrilling regions of thick-ribbed ice;' and, if they declared themselves no longer to be afflicted with water on the brain, their foes might have replied that ice on the brain was an equally dangerous complaint. The advocates of ice, however, were not all agreed. Some put their trust in floating ice, that is, icebergs. This was the first form of the craze. Lyell and Darwin believed in 'the great submergence.' This theory has of late years been largely abandoned, and land-ice in the form of glaciers is generally believed to have had more to do with forming the drift deposits than anything else. Doubtless it does explain a great deal—that must be fully admitted; but, after all, water appears to have had some share in producing the deposits in question. We must try and do justice to both sides.

It is always interesting to watch the growth of opinion on any subject, and to trace the gradual evolution of true theories out of false ones. And, should anyone remark that scientific men are constantly changing their opinions, therefore nobody need pay any attention to their theories, because the next generation of *savants* will say something quite different, and declare with the Frenchman, 'Nous avons changé tout cela,' then it may be replied that they are no more guilty in this respect than other people, philosophers, theologians, historians, politicians, and economists. Even in the last fifty years what advances have been witnessed, how many fallacies and false doctrines abandoned? We have often heard people speak mockingly of the changes that have taken place in scientific thought, and therefore feel bound to plead for geologists that, if blame there be, they are no more to blame than other people. And if, in early days, they were a little too prone to form theories before the facts were fully collected, that is a common weakness. The very men who scoff at scientists for changing have, as often as not, changed their own opinions, perhaps more than once. Let them remember that the discovery of truth is a slow and arduous process, and, in the very nature of the case, there must be evolution or change.

It will be worth while briefly to trace the course of opinion on this most difficult and intricate subject. There is no branch of the science so full of problems as what is called Pleistocene geology, that is, the geology of the last or Pleistocene period; in other words, the time of the Ice Age, so called, and the events immediately following it. One of the chief difficulties encountered in the endeavour to interpret drift deposits lies in the fact that, with the exception of moraines and a few gravels, &c., we cannot see such deposits actually in process of formation at the present day. It is not possible to penetrate very far under a glacier, much less to get below the Greenland ice-sheet. The more we can learn of the latter, the better shall we be able to understand some of the drift deposits.

Emmanuel Swedenborg,[1] who in his earlier days was assessor in the School of Mines in Sweden, was probably the first to describe the erratics so conspicuous in that country, and to endeavour to explain them. He put forward the theory that they had been deposited there by a great marine deluge; and thus he also accounts for some of the other phenomena, such as the peculiar ridges of sand and gravel known as Äsar (in Scotland as Kames). That great early observer, De Saussure, in his 'Voyages dans les Alpes' (1779), discusses the Jura erratics, and concludes that water alone could have brought them there, and at the same time formed the gravels that are found in the valleys. He believed that such water was in a state of violent action. He it was who first noticed those peculiar rounded and wave-like masses of rock to be seen in many a mountain valley known as *roches moutonnées*, and gave them that name, because they suggest a sheep's back.

---

[1] The following sketch is based upon the long and careful review of the whole history of the subject in Sir Henry Howorth's *Glacial Nightmare*.

We now know them to have been so rounded by glaciers creeping over them and wearing them down by means of their contained stones, but he attributed their shape to the action of water. In order to do this it was necessary to invoke the aid of one of those 'cataclysms' of which the early geologists had so much to say. It was a kind of colossal earthquake, producing a rush of water among other effects. This view was accepted by Hutton, the great founder of the sober modern idea of uniformity in geological action. That was in 1802. Also by his illustrious exponent, Playfair, though both of them afterwards abandoned the theory. But they did not invoke rushing waves; river water satisfied them. They thought that mountains were formerly much higher (which was quite true), and so a slope was provided for the waters to run down.

Sir James Hall, in 1812, also advocated a 'diluvial wave.' We pass on to the great Dean Buckland, whose pioneer work among the caves has been already alluded to. His 'Reliquiæ Diluvianæ' was published in 1823 with the object of showing that the Scripture account of the deluge was confirmed by science. Drift deposits, erratics, gravels, &c., were all supposed to be geological evidences of the Noachian Flood! Adam Sedgwick was at one time in the camp of the Diluvialists. Nor can the honoured name of De la Beche be omitted. Murchison, in 1835, expressed his belief in submergence and the dispersion of erratics by means of floating ice. But this marks the beginning of a new era in the controversy. Cataclysms and *débâcles* were no more talked of, but shore ice drifted about by winds and currents. This was a great advance. Darwin and Lyell came upon the scene advocating the cause of the new theory. Passing on to the final stage, we find that, according to Sir Henry Howorth, the first advocate of glaciers was Playfair, who had formerly been a Diluvialist. In his 'Illustrations of the Huttonian Theory' he alludes to a famous erratic at Neuchâtel called the 'Pierre-à-Bot.' It weighs 2,520 tons, and must have travelled seventy miles, being of granite. He argues that no current of water could have carried it uphill.

It is a pleasant surprise to notice that one or two humble natives of Switzerland had been intelligent enough to discover for themselves that the Swiss glaciers had formerly extended to places much beyond their present boundaries. In 1815 Charpentier was told this by a well-known chamois hunter with whom he passed the night in a hut. The same writer also mentions a woodman who had come to the same conclusion by observing big stones lower down the valleys. In 1822 Venetz proved the former extension of glaciers as far as the Jura Mountains. This was for those times a startling discovery.

The philosopher Goethe thus alludes to the new theory in his famous novel 'Wilhelm Meister' (second ed. 1829, vol. ii. ch. x.): 'Finally, two or three quiet guests called to their aid a period of intense cold, and saw in their mind's eye glaciers descending from the highest mountains like gliding roads, far into the low country, upon which, as on an inclined plane, heavy primary blocks were slid forward and farther onwards. So that, at the period of thawing of the ice, they sank down and remained permanently on the foreign soil.' The famous M. de Charpentier (1835), after at first opposing the theory, became a strong adherent to it, and it was he who really established it on a firm basis. Agassiz now appears on the scene, and accepts the conclusions of Venetz and Charpentier. But, in truth, he got the theory from Karl Schimper, a botanist, who had actually discovered this truth for himself. He it was who found the famous glacier marks on the chalk of the Jura at Landeron.[1]

Speaking of Schimper's discovery, Sir Henry Howorth says: 'This fact he communicated to Agassiz, whose imagination was at length fired by the impulsive and picturesque conversations of his friend, *and he determined for the first time to appear in the character of a geologist*, and advertised a course of lectures to be delivered to his fellow-citizens of Neuchâtel.' For this purpose

[1] Charpentier published a memoir entitled *Notice sur la Cause probable du transport des blocs erratiques de la Suisse.*

he even borrowed from Schimper the notes he had used for lectures at Munich. Once only did he acknowledge his indebtedness to the botanist, and then only to some slight extent. And so the newspapers gave him all the credit, and poor Schimper was forgotten. So in this country the same thing happened, and we were always told in our youth that the discoverer of glacial action in the past among the mountains of Northern Europe was Agassiz. The term 'Ice Age' (German, *Eiszeit*) was actually invented by Schimper.[1] This is a cunning form of theft which, we are sorry to say, is still rather prevalent even at the present time !

Dean Buckland was won over by the arguments of the brilliant naturalist. Murchison refused to be convinced by a paper read by Buckland to the Geological Society in 1840. An animated discussion took place, in which Murchison sarcastically remarked : 'The day will come when Highgate Hill will be regarded as the seat of a glacier, and Hyde Park and Belgrave Square will be the scene of its influence.' Whewell, the great Master of Trinity College, also protested.

However, Agassiz certainly helped Charpentier to convert others, both by his enthusiasm and his eloquence. 'Every terminal moraine,' he said, 'is the retreating footprint of some glacier, as it slowly yielded possession to the plains and betook itself to the mountains. Wherever we find one of these ancient semicircular walls of unusual size, there we may be sure the glacier resolutely set its icy foot, disputing the ground inch by inch, while heat and cold strove for the mastery. . . . By these semicircular walls we can trace the retreat of the ice as it withdrew from the plain of Switzerland to the fastnesses of the Alps. It paused at Berne, and laid the foundations of the present city, which is built on an ancient moraine.'[2]

---

[1] The responsibility for the above statements rests with Sir Henry Howorth, whom the author quotes.
[2] For a brief account of the work of ice, see the author's *Story of the Hills*.

But to return to our own country, the only theory which seems to account at all satisfactorily for the glacial deposits of Britain, at least as far south as the Midlands, is that of a modest local or British ice-sheet formed by confluent glaciers, whereby Scotland, Wales, and the North of England, besides much of Ireland, were to a large extent buried in snowfields and glaciers. What are shallow sounds round our coasts, were then mostly valleys filled with ice. Thus Scotch boulders were brought down to Cheshire or the North of England, as the case might be. Till accumulated at certain places under the old glaciers, and, as at last they slowly melted away, it was left behind; while the last small remnants of the old glaciers, before all the melting was finally accomplished, produced some of those very fresh-looking moraines that one sees high up among the mountains of Wales or Scotland. Much good careful work has been done in mapping boulders in the North of England, and in this way the dispersion of granite blocks from Shap, in the Lake District, has been worked out with very curious results; so much so, that some geologists still believe that only floating ice could have accomplished the work. South of the Thames we meet with no more boulder clays or moraines, but a peculiar set of drift deposits alluded to in Chapter I., which must have been largely formed by the action of water and melting ice. Doubtless every summer a great amount of ice was set free in the form of water, and flowed away southwards.

Now this is a very different conception to the Polar Ice-cap theory of Agassiz, Croll, and Professor J. Geikie. If the ice which once filled our mountain valleys came all the way from Arctic regions, how is it that boulders from those high latitudes have not been detected? Sir Archibald Geikie [1] expressly says that in the Scotch till he has never found stones which must have come from a long way off, but always such as might have come from those at no very great distance. It is clear we must give up

[1] *Scenery of Scotland*, p. 365 (2nd edit. 1887).

the idea of a great ice sheet having buried up all the whole country, except a few of the higher peaks. And instead we must think of the so-called Ice Age as a period of local and more or less coalescent glaciers. How else could moraines have formed, for they are the *débris* of the higher parts of the mountains that remain uncovered by snow? Recent Danish explorations have shown that even Greenland is not simply a sea of ice, as we have been so often told by the extreme glacialists, with nothing but snow and ice stretching away for leagues. It turns out to be a plateau, with many a bare rocky prominence exposed to atmospheric influences.[1]

Ever since the days of Sedgwick, and even before, the presence of boulders of Scandinavian rock at certain places on our eastern coast has been recognised. They have also been noticed in Russia and Poland. Believers in the great ice sheet naturally bring these facts forward as evidence for their theory, and would have people believe that such were carried for hundreds of miles east and west on the surface of slowly-creeping ice. But there are many difficulties in this view. In the first place, as a mere question of physics, unless we grant a very great upheaval of the Scandinavian peninsula, where is the slope required to cause the ice sheet to flow? We must have a slope of 3 or 4 degrees, or more, to make ice move, and even then it wants connection with moving masses higher up to urge it on. Secondly, the Norwegian geologists have found that the Loffoden[2] Islands do not show signs of glaciation right away up to their summits, as they ought certainly to do if they were once buried deep beneath an ice sheet ! Nor do the higher Scotch Mountains. Professor James Geikie, in his 'Great Ice Age' (1894), gives a coloured map showing its huge dimensions, according to his theory. This is chiefly based on the general directions taken by the glacial 'striæ'

[1] See the illustrations in the Danish Investigations in Greenland (*Meddelelser om Grønland*), undertaken by a commission.
[2] Howorth, *Geol. Mag.*, vol. i. decade iv., 1894, p. 496.

or scratches, which have been carefully mapped out by a large number of geologists. He and others believe that it met and turned aside the Scotch sheet, and, being deflected somewhat south, left the Scandinavian boulders on the Yorkshire coast. It is, to say the least, difficult to accept such a conclusion. Others hold that the stones drifted here on icebergs.

Sir Henry Howorth has ingeniously tried to show that they may have been brought by our Viking invaders as ballast for their ships, and in some cases left on or near the shore by shipwrecks. But it seems that they are too numerous for this explanation, and also are not merely found on the shore, but actually embedded in the boulder clay.[1] The same writer says: 'The Scandinavian ice sheet was, I believe, the invention of Croll, who, sitting in his arm-chair, and endowed with a brilliant imagination, imposed upon sober science this extraordinary postulate. He did not dream of testing it by an examination of the coasts of Norway, or even of Britain, but put it forward apparently as a magnificent deduction. Thus it came about that the great monster which is said to have come from Norway, goodness knows by what mechanical process, speedily dissolved away on the application of inductive methods. Of course, it still maintained its hold upon that section of geologists who dogmatise in print a great deal about the Glacial period before they have ever seen a glacier at work at all; but I am speaking of those who have studied the problem inductively. First, Mr. James Geikie, a disciple of Croll, was obliged to confess that this ice sheet, which is said actually to have advanced as far as the hundred-fathom line in the Atlantic, and there presented a cliff of ice like the Antarctic continent, never can have reached the Faroes, which have an ice sheet of their own. Next, Messrs. Horne and Peach were constrained to admit that no traces of it of any kind occur in the Orkneys or in Eastern Scotland.'

[1] On reading this Sir Henry writes: 'I agree with you here, and was mistaken in attributing these stones to the Vikings.'

## THE MYTH OF THE GREAT ICE SHEET 99

American geologists have for many years been carefully studying the effects of the Glacial period in North America, and the results arrived at there tend to confirm those obtained in Northern Europe. It is interesting to note that at the same period in the world's history there appears undoubtedly to have been a greater extension of glaciers in the southern hemisphere. In South America and in New Zealand there are evident signs of former glacial action where glaciers now no longer exist. Agassiz did not hesitate to express the opinion that even in tropical regions his colossal ice sheet had left its mark, but that was the result of his imagination!

As the reader will readily perceive, our object in introducing these geological topics is to endeavour to show what were the conditions of Europe when the race of Palæolithic hunters lived These men, as already stated, are believed by many geologists to have been here during the Ice Age. Others, however, refuse to accept this conclusion, and assert that man was *post*-Glacial. Many consider that the presence of the mammoth, and other extinct quadrupeds that loved cold, seems to show that the former opinion is the true one.[1] It is not necessary, however, as has been already pointed out, to suppose that the southern mammals found the climate too cold for them. Quite the contrary, for there appears to be abundant evidence to show that both were living in Europe actually at the same time! If in the Alps the fauna and flora of the snow-fields are different from those of the rich warm valleys below, how much greater must the difference have been in the Glacial period, either there or in Britain?

The mind is all the more vividly impressed with the majesty

---

[1] Mr. S. B. J. Skertchly, in 1878, considered that he had found Palæolithic implements *under a glacial deposit* at Brandon (Norfolk). But according to Professors Hughes and Bonney, with whom the writer visited the spot, they appeared to have slipped down a hole or pit in the deposit, and so were *not in situ*. Nor were they associated with extinct mammalian remains. However, some geologists accept them as evidence of man's inter-Glacial age (see Chap. VII.).

of these rivers of ice, as they have been termed, when the surrounding vegetation is green and luxuriant. Some of the most beautiful glaciers in the Alps descend right down into the midst of forests of firs, beeches, and larches; and it is through the green foliage of the trees that we catch a glimpse of the white waves of the icy sea and the dark walls of the moraines. In other places, fields of corn, or even vineyards and gardens, extend to the very base of the solid river, and sometimes, it is said, people have to mount upon fallen blocks of ice in order to gather the fruit off the branches of the cherry trees. Thus the vegetation of the temperate zone and the ice-fields of the pole, which on the Continent are separated from each other by thousands of miles, are here brought side by side. In the face of these facts, the mind rebels against the theory of warm inter-glacial climates advocated by so many geologists. It is far more reasonable to suppose that the intercalated sands, gravels, &c., with their peat, trunks of trees, and other organic remains, were caught up by the advancing glaciers, and covered over with till. In some cases they might even contain pre-Glacial forms of life; for in the early part of the period the glaciers must have gradually advanced over ground covered with Pliocene forms of plant and animal life. And just as much of the till is in reality old Pliocene soil, so much of the material thus intercalated with it must have accumulated in the previous period. It could not have vanished suddenly on the first appearance of the ice!

How much simpler and more intelligible this explanation is than the elaborate and complicated scheme of Professor James Geikie, given in his book,[1] we leave it to the reader to judge. Those who are not tied down to all the absurdities of the astronomical theory will find it difficult to believe that there were *six separate Glacial epochs, and five warm inter-Glacial ones!*

Whatever may be said against the diluvial theories of the old geologists, and their ingenious attempts to show that their belief

[1] *The Great Ice Age* (1894), p. 607.

in a universal deluge was supported by geological evidence, yet one cannot help feeling that a careful and unbiassed review of all the facts does rather lead one to the opinion that, after all, there was at least some truth in their ideas, and that, although they were absurdly exaggerated, such were *partly* justified. Only, instead of finding in geological records any proof of a universal catastrophe, we see only the signs of *great local floods*. These, however, were of wide extent, and their deposits in some cases are spread about over large areas independently of the present features of the ground. They are the plateau gravels, loess (including under that term the loams and brick-earths), and the high-level gravels. Every geologist, however strongly he may hold the doctrine of uniformity as taught by Lyell, will yet find it impossible to resist the conclusion, that about the end of the Glacial period inundations took place on a scale to which we can at present find no parallel. It is not therefore surprising that those of the old school took an extreme view of this evidence. Thus we find St. Pierre saying that the Deluge was caused by the simultaneous melting of two enormous polar ice-caps, the waters from which, in two great *débâcles* from north and south, overwhelmed all the low grounds of the world. Later on, a celebrated French geologist, M. Elie de Beaumont, could see no other way of accounting for the transport to places a long way off of erratics from the Alps, except by assuming that enormous currents of water flowed from them, which were derived from the melting of vast snows upon their lofty heights.

Geologists have long perceived that all the larger river-valleys of Europe, to say nothing of other regions, give evidence of having been subjected to floods on a much greater scale than anything to be witnessed in these days. Great deposits of alluvium, loam, brick-earth (known in Germany as 'loess' and in France as 'limon') can only be interpreted in that way. In various parts of Europe they spread over such large areas that some geologists hesitate to attribute them to river-action at all

But there is a great deal to be said for the view of Sir Joseph Prestwich, that they are part and parcel of the ancient river accumulations of the Pleistocene period, and were formed at the same time as the old gravels of which we spoke in Chapter I. The loess is a fine loam, consisting of a mixture of clay and carbonate of lime, and very fine-grained. It is best seen in the regions watered by the Rhine and its tributaries. There it rises from the margins of the alluvial flats at the bottoms of the valleys to a height of 200 to 300 feet, sweeping up the slopes of the valleys, and imparting to many districts a rich fertility. It is of greater thickness in the valleys and thinner on the higher slopes and plateaux, and lies like a great winding-sheet over the country. Land shells, stems of trees, human remains, and the bones of many mammals, both extinct and modern, have been found therein. Near Zeiselberg, at the mouth of the Kamp Valley, a great many bones were found under a black layer of charcoal and worked flints, which represents an old camping ground.

These and other remains [1] testified to the presence here of hunters of the Older Stone Age, together with the mammoth, reindeer, rhinoceros, &c. Many are the theories that have been put forward to account for this puzzling formation. That of Central Europe was at one time supposed to be of marine origin, but this theory has been abandoned, although two recent writers have endeavoured to show that the loess of China was formed under the sea.

Others, with Hibbert, endeavoured to prove that in the Rhine Valley it had been laid down quietly in a great fresh-water lake occupying the broad, open part of the valley above Bingen. But this lake theory had also to be abandoned. Guembel believed that at the end of the Glacial period a sudden depression of the Alps took place whereby all their glaciers were melted, and that such rapid melting of snowfields and glaciers produced vast floods. He was nearer to the truth; but no sudden depression of

[1] Vide *The Great Ice Age* (1894), p. 662.

the Alps is necessary. The water thus set free was supposed to have collected in a vast inland lake, in which the deposit was quietly formed. Sir Charles Lyell did not accept the lake theory, but he advocated the idea of a subsidence of the Alps, whereby the fall of the rivers was very much lessened, so that they took to depositing a fine silt. Others, with Mr. Belt and Sir A. Geikie, have suggested that the rivers of Central Europe had their waters dammed back by the advance of the great northern ice sheet, forming a *mer de glace* in the North Sea. But this will not apply to many other regions, where the loess is seen on a great scale, and therefore is unsatisfactory.

Baron Richthofen, in his great work on China, advocated a new theory, to the effect that in that country the loess was a superficial accumulation due to the long-continued action of wind piling up sand deposits. On the Yellow River it forms cliffs as much as 500 feet high. The true theory, at least for Europe, would appear to be that stated by Sir Joseph Prestwich years ago, to the effect that all the loams, under whatever name they may be known, are the flood deposits of the rivers formed at a time when they flowed at much higher levels, and *not* after the valleys had been cut down so deep as we now see them. They are therefore of the same age as the higher gravels, and really belong to them. They are the flood loams of the Glacial period, or at least of the close of it. Some writers even speak of a 'Pluvial period;' but we think that such a term may be misleading, and would condemn it on the same ground as we condemn the expression 'Ice Age.'

It is somewhat remarkable to find a distinguished geologist like Sir Joseph Prestwich harking back to the old catastrophic notions, and advocating a sudden submergence and consequent wave of diluvial waters spreading devastation over the face of the land, but in some recent papers he has advocated a theory of this sort. Sir Henry Howorth also has laboured hard to show that the diluvialists were right.[1]

[1] *A Possible Cause of the Origin and Tradition of the Flood*, by Sir

Among the phenomena to which the former would apply his theory are what is termed the 'head' or angular rubble noticed as overlying the raised beaches at some places on our south coast (it is a kind of talus, the rubble drift, a residue of the ordinary drift deposits which, according to him, did not appear to admit of any of the usual explanations), the high-level loess, and lastly certain breccias containing many bones of animals in fissures and elsewhere. These phenomena he traces over various parts of Europe, and, laying aside all the other theories, he puts forth his own, in which he boldly suggests that in Pleistocene times, in consequence of some submarine upheaval, the sea rose and submerged the whole of Western Europe and the Mediterranean coasts for a *short time*, leaving only some of the higher peaks and mountains dry.

The great number of bones, often broken, which have been found in the fissures in limestone at Gibraltar, and in caves in Sicily and elsewhere, certainly is remarkable; and Prestwich's theory is that the animals fled up the mountains from the invading marine flood, and either fell over precipices or were drowned in caves, &c. As the land rapidly emerged, the water in rushing off swept down the *débris* which had previously accumulated, and thus the rubble drift, &c., were formed and the bones of dead animals were swept into fissures. It is a bold theory, but one which most geologists are unwilling to accept. It asks too much for these days, when *débâcles* and waves of translation are no longer fashionable, and when it is generally agreed that the ordinary phenomena of Nature are found to be sufficient to explain all that is to be found in the geological record. He thinks that the tradition of the Flood would never have arisen but for some event of a very exceptional and extraordinary character —something much greater than an ordinary river flood. And so,

Joseph Prestwich (Trans. Victoria Institute), vol. xxvii. p. 263; and *The Tradition of the Flood* (London: Macmillan & Co. 1895), and *Quart. Journ. Geol. Soc.*, vol. xlviii. p. 326; Howorth, *The Glacial Nightmare and the Flood*.

not seeing his way to explain the facts by the usual agencies of which the action is known, he invokes a theory of a rapid submergence and emergence. Such, he thinks, may well agree with the magnitude of the catastrophe as recorded in Scripture. A deep impression was made thereby on the human mind, because it was considered as miraculous and due to the anger of the gods. Hence the allegorical accompaniments. One argument which, to our mind, is fatal to this theory, is that both the account in Genesis and that in the primitive Accadian or Chaldean version mention *rain* as the cause of the Flood.[1]

Sir Henry Howorth has endeavoured to show that the sweeping away of the mammoth and its contemporaries can only be explained by a flood. In his later book he applies the same theory to the drift deposits : 'Water moving in this way would sweep up and mix together, and then throw down in great heaps, when its current was arrested, all the *débris* that came in its way, and it must have been no ordinary marine submergence, but a widespread wave of waters which would pass over the country mixing the materials it met with. In the case of such a flood, again, we should have the materials of the loose covering dominated very largely by the nature of the subjacent strata over wide districts.'[2]

It would be quite impossible to give in the course of a single chapter any account of the peculiar drift deposits. To do that would require a large volume ; our object has been to show in the first place that the Ice Age has been greatly exaggerated, and the notion of a Polar ice-cap must be for ever abandoned ; secondly, to give a little sketch of the history of opinion on the subject, and incidentally to show that at the end of the Glacial period water, resulting from the melting of so much ice, played a certain part in making the drift deposits ; and lastly, to give some idea of the

---

[1] See Professor Sayce, *Fresh Light from the Ancient Monuments*, 1892, p. 28.

[2] *The Glacial Nightmare and the Flood.*

condition of Europe at the time when it was inhabited by the Palæolithic hunters, that is, assuming, as we think we are justified in doing, that they *did* live in the Glacial period. And with regard to the Flood, we are inclined to agree with Sir Henry Howorth and Mr. W. Shone and some others, that the *great gap* so generally recognised between the Older and the Newer Stone Ages may have been due to the vast physical changes that took place at the end of the Glacial period, the amelioration of the climate, and consequent melting of snow and ice causing great floods in river-valleys, and the extinction or the migration of many animals. French writers apparently recognise no interval between the Palæolithic and Neolithic periods. It remains for future workers to prove whether man was *pre*-Glacial, *inter*- or *post*-Glacial.

In doing so we have endeavoured to avoid details such as would only weary the reader, and it is to be hoped that, after perusing our remarks, he will not share the feelings of a schoolboy who wrote in his Euclid the following lines :

> If there should be another flood,
> Hither for refuge fly ;
> Were the whole world to be submerged
> This book would still be dry !

## CHAPTER V

CHANGES OF CLIMATE AND THEIR CAUSES

*Not from the stars do I my judgment pluck;*
And yet methinks I have astronomy,
But not to tell of good or evil luck,
Of plagues, of dearths, or season's quality;
Nor can I fortune to brief minutes tell,
Pointing to each his thunder, rain, and wind,
Or say with princes if it shall go well,
By oft predict that I in heaven find.
SHAKESPEARE, *Sonnet XIV.*

EVERY geologist is well aware that in past ages the climate of particular regions has undergone considerable change. Palm trees once flourished where the Thames now flows leisurely along; and yet, at a later epoch, the Arctic willow grew in the same district, and the musk-sheep browsed amid winter snows! And, more wonderful still, when we come to examine the strata in Arctic regions, we find they contain fossils which give undoubted evidence of genial or even tropical climate. It has been discovered of late years that during at least three former geological periods plants and animals have abundantly flourished in very high latitudes where no forest tree can live at the present day! Such discoveries, made independently by different travellers, are among the most remarkable results of geological research, and have given rise, as we shall see, to a variety of ingenious speculations.

But first, let us have a few more facts. In North Greenland a flora has been discovered extending at least up to latitude 70°, which, according to Heer, contains over thirty species of conifers, besides beeches, oaks, planes, poplars, maples, walnuts, limes, magnolias, and many more. There can be no doubt that these grew on the spot, instead of being transported by rivers, or in any other way, for their fruits in various stages of growth have been found with them. They are of Miocene Age, a period characterised by a warm climate. No fewer than 136 species of fossil plants from Spitzbergen (lat. 78°) have been described by Heer.

A report by Captain Fielden, Mr. De Rance, and Prof. Heer says that the Grinnell Land lignite indicates a thick peat moss, with probably a small lake, with water-lilies on the surface of the water, and reeds on the edges, with birches, poplars, and taxodiums on the banks, and with pines, firs, spruce, elms, and hazel bushes on the neighbouring hills. Now at the present time the taxodium is confined to Mexico and the south of the States. Such a sylvan landscape as this seems entirely out of place within 600 miles of the Pole, to which these Arctic forests must have extended, if land reached so far. An English Arctic expedition, not many years ago, discovered a thick bed of coal in lat. 81° 45′ from 25 to 30 feet thick, and full of land plants, especially conifers. The Parry Islands, in lat. 76° 20′, have yielded tropical shells of the Jurassic period. In the Triassic beds of Spitzbergen certain well-known mollusca have been recognised, such as ammonites, and the delicate nautilus, which we all know prefers warm water to cold. Alaska and the Mackenzie River have likewise yielded remains of plants of Miocene Age, indicating a climate such as is now enjoyed by the northern parts of Italy.

The late Prof. Haughton claimed to have shown that the Tertiary plant remains, indicating a climate similar to that of Lombardy, are so situated round the North Pole that no possible change in the position of that pole (even were such permitted by mechanical considerations) would give them the warm climate which they require. But of this supposed cause we shall have more to say presently.

When scientific men and others began to speculate on the reason of such remarkable changes in climate as are indicated by the facts alluded to above, the first idea which occurred to them was that in the slow cooling of the globe a true cause might be found. It was not unnatural; astronomy, according to the nebular theory, tells of a time when our planet was a molten globe; it must therefore have been for ages cooling down. Why, then, should not part of such cooling have taken place in former geological ages? Other facts known to geologists encouraged the idea; for instance, the wide distribution of the fossil forests of the Carboniferous period, and the apparent absence of zoological provinces in remote periods. But it seemed to be forgotten that subterranean heat has very little effect on the surface, at least judging from the case of a volcano. Many active volcanoes have a permanent covering of snow, and it is possible, though dangerous, to walk on the thin crust of a lava stream with molten rock only a few inches below you. Lord Kelvin says: 'The earth might be a globe of white-hot iron covered with a crust of rock about 200 feet thick, or there might be an ice-cold temperature everywhere within thirty feet of the surface, yet the climate would not on that account be sensibly different from what it is, or the soil be sensibly more or less genial than it is for the roots of trees or smaller plants. Yet greater underground heat is the hypothesis which has been most complacently dealt with by geologists to account for the warmer climates of ancient times.'[1]

Others, finding no help in the earth beneath or in the waters under the earth, or the supposed seas of lava there, took refuge in the atmosphere, and suggested help from that quarter. It was surmised that, perhaps, in ages past the state of our atmosphere was different. It might have been more capable of keeping in the earth's heat, or it may have been thicker, that is, deeper. But supposing that at one time it contained more aqueous vapour, that would to some extent tend to keep the earth warm by stopping radiation outwards, but, through the action of clouds, it

[1] *Quart. Journ. Geol. Soc.*, viii. 58.

would also keep out more of the sun's radiant heat. And so this change would cut both ways.

It is a mistake to suppose that in the past changes of climate have always been in one direction, and that for the worse. Those who appealed to the cooling of our planet either forgot or were not aware of the great improvement in the climate which has taken place in Europe since the Glacial period.

Others, soaring into dim regions, have even suggested that the sun, and with it the whole solar system, may at one time, *i.e.* during the Glacial period, have been traversing through some colder region of space. That the sun has a 'proper motion' of its own is not denied; but we know absolutely nothing about the regions of space, and therefore to assume that some are warmer and others colder is, to say at least, quite unscientific.

Erman and Babinet have suggested a modification of this theory; they argue that possibly some regions through which our system has formerly passed may have been more thickly strewn with meteorites and meteoric dust than others; and so, just as our earth *apparently* suffers to a slight extent by passing through a band of meteorites in the early part of the month of May, so a general lowering of temperature may have taken place in the Pleistocene period by our passing through a region of space thickly strewn with meteors.

Comets we know are connected with meteorites, and, in one case at least, a broken-up comet appears to have turned into a band of meteors; but it would not take the earth very long to pass through an ordinary comet, and we have no right to assume the existence of vast masses of meteoric matter of which we have absolutely no knowledge. But a still better objection to all such theories lies in the simple fact, now pretty well recognised, that the Glacial period was, after all, only a local phenomenon like the flood. New Zealand has *not* passed through a Glacial period.

Of late years astronomers have discovered that certain stars go through very great variations in size and brilliancy. A few years ago a certain star was said to be blazing out in a prodigious way, and in a short time it resumed its former size. No satisfactory explanation could be offered of this strange sight. Had it collided with some other star, and so received a sudden accession of heat? Or, had this any connection with the well-known fact that some stars, like one in the constellation Cetus, go through regular but short periods during which their brilliancy varies? This regular change is supposed to be due to the presence of dark bodies like planets revolving round them, and so at times partially eclipsing them.

Again, sun spots have been appealed to. At times the solar atmosphere appears to be unusually active, and at such times we see more spots on the disc. These periods of special activity appear to come once in about eleven years, or nine times in a century. They are believed to affect the weather,[1] and so why should they not have affected climate in former ages by undergoing still greater variation? Here, again, we must answer that all astronomical theories fail to meet the case of the Glacial period because it did not affect the whole world. Had this important fact been sooner recognised, a great deal of unnecessary speculation and argument would have been saved. But, unfortunately, geologists were misled by Agassiz and others, who thought they saw signs of Glacial periods everywhere! Was not the wish father to the thought?

Turning once more to the sun, we find other writers suggesting that, as in the course of ages the sun kept on cooling and therefore contracting, till it attained something like its present size, changes must have taken place in the amount of heat received, and in the length of a summer or a winter day, and so the climate of the world generally may have become less genial. In the later history of the solar system the sun may have been much less dense, and so have occupied a bigger space in the centre

[1] See Lockyer's *Astronomy*.

of the system. Is it possible that this was the case in the Miocene period? Then it might have caused much more illumination of the polar regions. The idea is at least worth considering. Most of these theories have one radical defect which has been forcibly pointed out by the late Prof. Tyndall—namely, that they assume great cold to be a necessary condition for the production of a Glacial period. It is not cold alone that is wanted to produce glaciers. You may have, as we all know, the bitterest north-east winds without a single flake of snow. But what *is* a necessary condition is the precipitation of aqueous vapour in the form of snow. Let us hear what this great exponent of physical science says: 'It is perfectly manifest that by weakening the sun's action, either through a defect of emission, or by the steeping of the entire solar system in space of a low temperature, we should be cutting off the glaciers at their source. Vast masses of mountain ice indicate infallibly the existence of commensurate masses of atmospheric vapour, and a proportionately vast action on the part of the sun. In a distilling apparatus, if you required to augment the quantity distilled, you would not surely attempt to obtain the low temperature necessary to condensation by taking the fire from under your boiler; but this, if I understand them aright, is what has been done by those philosophers who have sought to produce the ancient glaciers by diminishing the sun's heat. It is quite manifest that the thing most needed to produce the glaciers is *an improved condenser*; we cannot afford to lose an iota of solar action; we need, if anything, more vapour, but we need a condenser so powerful that this vapour, instead of falling in liquid showers to the earth, shall be so far reduced in temperature as to descend in snow.'[1]

If, then, we can find no satisfactory and efficient cause, either in changes in the internal heat of the earth, or in its atmosphere, nor in the sun itself, nor in the regions of space, let us look to the earth's axis of rotation and see if any change in the position of that

[1] *Heat a Mode of Motion* (1888), p. 189.

will help us in explaining former changes of climate. It is, of course, well known that a slight inclination of the earth's axis, namely $23\frac{1}{2}°$, towards the ecliptic (or plane in which the earth revolves round the sun) is the cause of our seasons. Were the axis perfectly upright, *i.e.* at right angles to this plane, we should have no seasons at all, but equal day and night everywhere. Again, were the axis actually to be in the ecliptic itself, we should have perpetual day on the hemisphere turned towards the sun, and perpetual night in the other one. As it is, both these undesirable extremes are avoided. Hence astronomers have speculated whether the earth may not have been shifted as a whole, so that what is called the 'Obliquity of the Ecliptic' varied. Now, it does actually vary to some extent, but only very slightly. According to Laplace, the greatest amount possible is $24°\ 35'\ 58''$, and the least $21°\ 58'\ 36''$. It varies about $48''$ per century, owing to the attractions of the planets. These are very narrow limits, and it is clear that such variations could never have brought about any considerable change in climate.

It has been a favourite speculation with some that in former times, by a shifting of the earth's axis of rotation, the poles may have wandered to different points on the surface of the globe. The notion seems to be as old as the fifteenth century. Laplace and others, however, have stoutly denied the possibility of such a change ; but let us see what others have to say. In the early stages of the world's history, before oceans or continents were in existence, and when the crust was less rigid, it is quite conceivable that considerable changes of that kind may have taken place. This has been admitted by two eminent authorities, Lord Kelvin and Professor G. Darwin. The former conceives that in such times the axis may have shifted 30° or 40° or more. But, unfortunately, this will not help us, because we are dealing with the *latest* geological period. When once the earth had become as rigid as it is now, it is hardly conceivable that the position of the poles could have shifted much. Nothing short of truly enormous displacements of matter on the surface, producing great geographical changes, could cause any considerable wandering of the poles. Already there is a large excess of solid matter round about the equatorial region; for, as the reader is probably aware, the polar diameter is shorter by about twenty-six and a half miles than the equatorial, thereby producing a great equatorial protuberance, having a thickness on each side of $13\frac{1}{4}$ miles. It is quite clear, then, that changes of corresponding magnitude are necessary to produce the required shifting of the poles. But the answer to the question partly depends on the theory that the interior of the earth is all solid ; *if* that is not the case, and *if* other conditions were satisfied, it is possible, according to Professor G. Darwin, that the poles may have wandered some ten or fifteen degrees. But these ' ifs ' produce great uncertainty.

Many geologists of the present day believe that, broadly speaking, the general distribution of land and water on the globe has been the same all through geological time. This idea, known as the doctrine of the permanence of ocean basins (and continents), has certainly received some support from the results of deep-sea exploration, especially from the work of the scientific cruise of the 'Challenger,' now fully recorded in twenty large volumes published by the Government. It has been rendered clear by these researches that the stratified rocks, as far as we know them, do not contain any of those peculiar and characteristic deep-sea deposits known as red clay, diatom ooze, radiolarian ooze, &c. The chalk was not formed in a deep sea, nor was the carboniferous limestone, nor, indeed, any other formation that we know of. Hence, although we must all admit that minor revolutions in geography have frequently taken place, otherwise our English stratified rocks could not have been formed under the sea, as we know most of them were formed, yet we cannot admit such vast revolutions to have taken place as would bring the deepest recesses of the ocean to the condition of dry land, or *vice versa*. But, on purely mechanical grounds, it has been shown that only vast elevations and depressions of the earth's surface could possibly bring about any shifting of the axis such as might cause the climate of the polar regions or of the temperate latitudes to have suffered any considerable change.

Our friend, the Rev. E. Hill, says: 'Mathematicians may seem to geologists almost churlish in their unwillingness to admit a change in the earth's axis. Geologists scarcely know how much is involved in what they ask. They do not seem to realise the vastness of the earth's size, or the enormous quantity of her motion. When a mass of matter is in rotation about an axis, it cannot be made to rotate about a new one except by external force. Internal changes cannot alter the axis, only the redistribution of the matter and motion about it. If the mass began to revolve about a new axis, every particle would begin to move in a new direction. What is to cause this? When a cannon-ball strikes an iron plate obliquely, the shock may deflect it in a new direction. The earth's equator is moving faster than a cannon-ball. Where is the force that could deflect every portion of it, and every particle of the earth into new directions of motion?'[1] Lord Kelvin has shown both by theory and by experimental demonstration that, for steady rotation, the axis round which the earth revolves must be a principal axis of inertia, and, to alter this, vast transpositions of matter at the surface, or else distortions of the whole mass, *must* take place to produce such a shifting of the poles—*i.e.* of the axis—as will cause any considerable change in climate at a given latitude.

Another very acute reasoner, the late Dr. Croll, says : 'There never was an upheaval of such magnitude in the history of our earth. And to produce a deflection of 3° 0′ 17″ (a deflection which would hardly sensibly affect climate) no less than one-tenth of the entire surface of the earth would require to be elevated to a height of 10,000 feet. A continent ten times the size of Europe

[1] *Geol. Mag.*, 1878, p. 265.

elevated two miles would do little more than bring London to the latitude of Edinburgh, or Edinburgh to the latitude of London. He must be a sanguine geologist indeed who can expect to account for the glaciation of this country, or for the former absence of ice around the poles, by this means. We know perfectly well that since the Glacial epoch there have been no changes in the physical geography of the earth sufficient to deflect the pole half a dozen miles, far less half a dozen degrees. It does not help the matter much to assume a distortion of the whole solid mass of the globe. This, it is true, would give a few degrees' additional deflection of the pole; but that such a distortion actually took place is more opposed to geology and physics than even the elevation of a continent ten times the size of Europe to a height of two miles.'

We must pass on now to the consideration of a theory which is generally called 'The Astronomical Theory of the Ice Age,' as if none of the other theories above mentioned had ever been in existence. It is one which has been accepted by many geologists as affording the only satisfactory solution of the problem. It was always wisely rejected by Lyell, and, as we shall see presently, breaks down like the rest, and fails to satisfy either the astronomer or the geologist. Our chief reason for introducing it is on account of its bearing on the important question of the antiquity of man. As already stated in a previous chapter, many geologists have blindly followed Mr. Croll in his attempt to calculate how many years have elapsed since the disappearance of the glaciers of the Ice Age. This, he supposed, ended about 80,000 years ago, at least. Now, considering that the Palæolithic hunters were *certainly* here during the Glacial period, we should have to admit, if we accepted Croll's calculations, that these men lived more than 80,000 years ago. This we stoutly refuse to do. Can anyone who seriously examines the evidence bring himself to believe that such an enormous period of time has elapsed since Wookey Hole or Kent's Cavern was inhabited by men? It is ridiculous to suppose that the glacial striations in Wales or Scotland, to say nothing of moraines, &c., could have survived for so long a period. The forces of denudation should have destroyed them long ago! We shall, therefore, be in a better position to consider the question of the antiquity of man without prejudice when we have dismissed the astronomical theory from our minds, and with it all the absurdities into which it has led many an unsuspecting geologist. This may perhaps be considered rather strong language, considering that the theory has been so generally accepted, and that it still is believed to have the support of Sir Robert Ball.[1] But, on the other hand, it is fast being abandoned, and at last year's meeting of the British Association appears to have been successfully demolished by Mr. E. P. Culverwell. It has done a great deal of harm to both geology and archæology, and we may therefore rejoice in its overthrow. Professor G. Darwin, of Cambridge, has now ceased to support the theory.[2] We will first briefly state the theory itself, and then endeavour to show how it fails.

The earth is not always equally distant from the sun; we are, as a matter of fact, actually further from the central luminary in summer than we are in winter. The earth, in its annual revolution round the sun, describes not a circle but an ellipse, with the sun in one of the foci of the ellipse. We in the Northern Hemisphere have our winter in perihelion—*i.e.* when we are nearer to the sun—and our summer in aphelion, when we are further away. The mean distance of the sun is 92,400,000 miles, and the difference in distance at perihelion and at aphelion is now 3,000,000 miles. The present eccentricity is ·0168 of the earth's distance during the northern winter. This is only a very small fraction of the whole distance, so that it is necessary to draw the orbit on rather a large scale in order to show that it is not a true circle. The sun is out of the centre, and hence the orbit is

[1] See his book, *The Cause of an Ice Age* (1891).
[2] See recent correspondence in *Nature*, Sir H. Howorth and others.

said to be 'eccentric.' But, although we in the Northern Hemisphere are nearer to the sun in winter, the earth's velocity at that part of the orbit is, for that very reason, increased, thus making our winter shorter by nearly eight days than our summer. In the Southern Hemisphere, on the other hand, the winter is longer by the same amount. At the present time the earth receives in the perihelion part of its course one-fiftieth more heat than it does in the aphelion part in a given time. The total amount received annually in either hemisphere does not vary whatever the eccentricity, but its distribution between summer and winter does, because, the greater the eccentricity, the greater the difference in the length of summer and winter.

As things are now, the effect is very slight, but the astronomers have calculated that the earth's orbit from time to time varies considerably in shape, being sometimes much more eccentric than at other times, and with a high eccentricity the relative lengths of summer and winter will be very different. The cause of these changes is to be sought in the attractions of the larger planets, especially Jupiter and Saturn, on the earth. Now, they calculate that when these planets succeed in pulling out the shape of the earth's orbit as far as they possibly can, so that the difference between the two diameters of the ellipse will be as much as 14,368,200 miles, or ·07775 of the major axis, the difference in the number of days of summer and of winter will be thirty-six days instead of nearly eight. The orbit of the earth is slowly becoming more circular now, and in 23,980 years from A.D. 1800 it will be as nearly circular as it can be, with an eccentricity of only about 500,000 miles. Then it will slowly get more elliptical, until the limit above mentioned is reached. This was worked out by Legrange towards the end of last century, and more exactly by Leverrier in 1839.

Dr. Croll extended the calculations of Leverrier and computed the periods of greatest and of least eccentricity for the last 3,000,000 years. The results he arrived at indicated that within that long space of time there had been four or five principal periods of high eccentricity, with a few others in between when it was not so great, though more than now, and likewise corresponding periods of least eccentricity. But they occurred in an irregular manner. About 2,650,000 years ago it was at its minimum, but a period of great eccentricity occurred about 980,000 to 720,000 years ago; and again, another such period began about 240,000 years ago, and ended at 80,000 years ago. He referred the Glacial period to the latter of these. But, since, according to the theory, a Glacial period requires for either hemisphere that its winter be in aphelion, it is clear that glacial conditions could not have lasted all the time represented by the last two sets of dates. By precession the conditions would be reversed every 10,500 years. According to the above figures, the greater and longer period of eccentricity, which began 980,000 years ago, lasted 260,000 years, while the other, which began 240,000 years ago, only lasted 160,000 years. Now, if this theory and these calculations are correct, we ought to find evidence of an older and severer Glacial period. But geologists cannot find them. Other objections will be dealt with presently.

We said just now that the total amount of heat received by the earth annually on either hemisphere does not vary; what does vary is the amount received in a given time. This is in accordance with a simple theorem stated thus by Herschel, that the amount of heat received by the earth from the sun, while describing any part of its orbit, is proportional to the angle described round the sun's centre. It was this same great astronomer who, in 1830, pointed out that great variations in eccentricity might have very marked influence on climate. For, supposing, as now, our northern winter occurred in perihelion, we might have a very short but mild winter, the increased warmth being due to our nearness to the sun, while the shortness of that season would be due to increased velocity of revolution; and, on the other hand, a long cool winter, due to increased distance from the sun, and slow revolution over a much longer course. But the opposite hemisphere might be rendered uninhabitable by the fierce extremes caused by concentrating half the annual supply of heat into a summer of very short duration, and spreading the other half over a long dreary winter, sharpened to an intolerable intensity of frost when at its climax by the much greater remoteness of the sun. It is believed by the advocates of the Astronomical theory that in this way one hemisphere might be experiencing an Ice Age, while the other went through a sort of perpetual spring.

But this is not all; there are other slow changes taking place which must be briefly considered. These are the precession of the equinoxes and the revolution of the apsides, together with a small and much less important one known as nutation. The orange-like shape of the earth, with a difference between the polar and equatorial diameters of about twenty-six and a half miles, has already been alluded to. The equator, therefore, has a protuberant mass of matter with a thickness of some thirteen and a quarter miles; now the sun and moon together exert a pull on this mass, and thereby cause the axis of the earth to point to different parts of the heavens. In other words, the pole star varies, and our pole describes a circle.

We know from observations and measurements on the Pyramids that, in the time of the Pharaoh Cheops, who built the Great Pyramid, the earth's pole then pointed to a different part of the heavens. Therefore, the equinoxial points on our orbit keep slowly moving, hence the term 'precession of the equinoxes.' Were it not for another movement, this great cycle of change would be undergone in 25,868 years. But the same disturbing forces, namely, the pull of the sun and moon on the protuberant equatorial regions, brings about a gradual change in the direction of the major axis of the earth's orbit, called the motion of the aphelion, or revolution of the apsides. Now the combined effect of these two causes is to bring down the great cycle of change in the direction of the earth's axis in space to 21,000 years. In half that time, therefore, the axis has described half a circle—that is, it will point in just the opposite direction. This means that in either hemisphere the conditions will be reversed; thus we now have our northern winter in perihelion, but in the above space of time the pole will have worked round half a circle, so that the winter will then be in aphelion. Suppose, then, that such was the state of things now, but that, at the same time, we were living in a period of high eccentricity. It is argued that we should have a very long cold winter, with a short mild summer; and that in consequence so much snow and ice would form during such a winter that it could not possibly be all melted during the summer. We should, therefore, find ourselves passing through an Ice Age. Then in another 10,500 years the axis of the earth would have described a complete circle, and the same star would once more be the pole star. But by that time the condition of things would be reversed for the Southern Hemisphere, because the South Pole would be pointing away from the sun, and that region would be experiencing an Ice Age. Hence, by the Astronomical theory, each hemisphere would alternately experience an Ice Age, and a condition of climate resembling a kind of perpetual spring.

Sir John Herschel felt very doubtful whether anything like the effects here indicated as following from high eccentricity would actually occur. But Mr. Croll took up the subject and argued very cleverly, that with winter in aphelion a large amount of ice and snow would be formed, such as could not be melted during the short hot summer, to say nothing of other changes mentioned below. Hence from year to year, while these conditions lasted, an accumulation of snow and ice would take place in one hemisphere, with a kind of perpetual spring in the other. Then, after 10,500 years the state of things would be reversed.

Croll's arguments with regard to the indirect effects of high eccentricity, combined with winter in aphelion, are certainly very ingenious. In the first place, he points out that accumulated masses of snow and ice would tend to lower the summer temperature by direct radiation from surrounding bodies. Ice can never rise above 32° F. or 0° C. while it is ice, whatever may be the power of the sun as it shines on other things, for instance, on our faces. The sun's heat can only slowly melt the ice, and in so doing it is expended. But, snow and ice being good reflectors, much of the heat reaching the earth would be thrown back into space, and so lost to the earth. Thirdly, the cooling of the air by radiation would tend to produce thick fogs and cloud, and so to greatly diminish the supply of heat received from the sun. A great deal of aqueous vapour would be formed during the hot summer, but much of this would be condensed in the cool air into watery particles suspended in the air in the form of cloud and mist. There would in consequence be less melting of snow, so that snow and ice would fast accumulate.

It is well known that ocean currents have a powerful influence in modifying climates. We in Western and North-Western Europe owe a great deal to the Gulf Stream. Without that great source of warmth we should be subjected to winters as severe as are those of Canada. This great current brings large supplies of heat from the Gulf of Mexico and the equatorial parts of the Atlantic Ocean. Mr. Croll believed that the currents of the ocean are chiefly determined by the prevailing winds, and especially the trade-winds. These are certainly due to a difference of temperature between equatorial and polar regions, the warm air of the former rising up and going off to the poles, while the cold air from the latter comes creeping in along the surface of the earth in the opposite direction to supply its place. Thus, he argued that if the mean temperature of one hemisphere be lowered, the trade-winds will be all the more strongly attracted towards the pole of the other. So that during an Ice Age in the Northern Hemisphere the trades would blow more strongly away from it. Now, since the Gulf Stream is really an equatorial current of warm water flowing westward deflected in a north-easterly direction by the Gulf of Mexico and the American coast, it would follow that the winds blowing in the Northern Hemisphere from the north-east to the equator would be stronger than the corresponding winds in the Southern one, and consequently the warm equatorial waters would tend to be driven south instead of north, and so the Gulf Stream would be considerably diminished in volume. In like manner he argued that the upper trades to the north would be all the more powerful, and would thus bring to the northern and polar regions a larger amount of aqueous vapour to be condensed into snow. But, ingenious as are many of his arguments, there are not a few serious objections to the theory. These must now be briefly stated.

The first, and perhaps the most serious difficulty, is the fact that the formulæ relied upon by the astronomers cannot be used for very distant periods of time. The observations on which they are based do not go further back than 2,000 years at the most. We are not therefore justified in relying upon any calculations with regard to the eccentricity of the earth's orbit during the last 3,000,000 years, or even 80,000 years ago. Thus the very foundation of the theory on which such a big superstructure has been built is found to be weak.

But, (2) passing over this serious defect, and assuming that the calculations are to be relied upon, it would follow that during the last 3,000,000 years there were four or five periods of high eccentricity, each long enough to allow of winter occurring in aphelion (by precession of the equinoxes). But we ought to consider the whole of geological time and not merely a part of it. By geological time is meant the whole period, or rather series of periods, in the earth's history during which stratified rocks have been forming. It is not possible, unfortunately, to translate this into years, but it must represent many millions of years. That we are all agreed upon; and, considering that at least 100,000 feet of stratified rocks have been formed by slow processes in seas, rivers, and lakes during these periods, and that vast changes in the organic world also took place, including the evolution of fishes, reptiles, birds, and mammals, it is generally considered that at least 100,000,000 years were required to accomplish all this. That is the minimum accepted by geologists; but many would be inclined to accept the higher limit of Lord Kelvin—namely, 500,000,000 years. Taking the lower limit, however, there ought to have occurred 130 to 160 Glacial periods. Instead of that, some geologists have endeavoured to show that signs of glacial action occur at ten different geological periods. But the evidence, consisting of rounded and scratched boulders and breccias, is so unsatisfactory that the majority of these supposed cases of glacial action must be rejected. It is generally considered that there are genuine but very local signs of ice action in certain Permian deposits; also possibly in others of Eocene and Miocene Ages respectively. Now, it need hardly be said that, if the Astronomical theory be true, we ought to find signs of glacial action much more abundantly. Moreover, glaciated boulders in themselves are no sufficient proof of a Glacial period. It is far more likely that they merely prove the existence of local glaciers, like those of the Alps at the present time.

In the next place, (3) the glacial conditions of the Ice Age appear to have been limited to only one half of the Northern Hemisphere instead of extending all round it. There was no ice sheet over the greater part of Siberia and Alaska. It has even been argued that the climate of Arctic regions was actually *milder* during this period than it is now. Anyhow, there is no doubt but that during that time there was enough vegetation in Siberia to afford food for an enormous number of mammoths and other creatures, as we know from the thousands of mammoth tusks found there, in spite of the fact that for centuries there has been a large trade in mammoth ivory.

Again, (4) if the theory be true, we ought to find evidence of alternating glacial and warm periods in each hemisphere. Now, there is apparently no evidence of an Ice Age in New Zealand.

Prof. James Geikie and others have attempted to show (5) that

## CHANGES OF CLIMATE AND THEIR CAUSES

our British glacial deposits contain records in the shape of interstratified sands, gravels, peats, &c., of warm interglacial periods. But these phenomena can be more easily explained by local advances and retreatings of glaciers. The supposed warm intervening periods must be abandoned.

And, lastly, (6) the recentness of the Glacial period is becoming much more generally recognised, and many geologists failed to see how the striations, moraines, and *roches moutonnées* could have lasted for anything like the periods required by the Astronomical theory. One is inclined to think that delicate striations and polishings would have been destroyed by atmospheric influences within the space of twenty thousand years.

We have endeavoured to show that this famous theory is open to criticism from six different quarters; and it would appear that these converging lines of attack have been successful in driving it from the field of geological speculation, a result which we hail with satisfaction.

The question we have been considering—namely, the cause of the Glacial period—is one of the most difficult problems in the whole of geology; it is therefore wiser at present to consider it an open question. It may, however, be mentioned here that some geologists, following the lead of Sir Charles Lyell, believe that geographical changes may bring about such an alteration of climate. Dana, Le Conte, Wright, Jamieson, and others believed that the former elevation of Northern Europe and America had much to do with bringing on the change. There is ample evidence of such elevation in fjords and submarine continuations of river valleys. Fjords are simply submerged valleys, and there are plenty of such round our Scottish coast. The elevation of Scotch, Welsh, and Scandinavian mountain chains would certainly cause increased precipitation of snow, and consequently cooler summers. Prof. Bonney [1] has shown that the temperature of the Glacial period need not have been very low. The following state-

[1] *Nature*, 1891, p. 373.

ment[1] of the order of succession of the glacial deposits, according to Professor J. Geikie, given in ascending order, may be useful to the reader and especially to travellers, although geologists will not all agree with his interpretation of them:—

GLACIAL SUCCESSION IN BRITISH ISLES.

1. Weybourn Crag and Chillesford clay.
2. Forest bed of Cromer.


3. Lower boulder-clays and associated fluvio-glacial deposits.
4. Marine, fresh-water and terrestrial accumulations; basin of Moray Firth; basin of Irish Sea, Lanarkshire, Ayrshire, Edinburgh, &c.; Hessle gravels; Sussex beach deposits; Settle Cave, &c.
5. Upper boulder-clay and associated fluvio-glacial deposits.

6. Fresh-water alluvia underlying oldest peat-bogs; probably a considerable proportion of our so-called 'post-glacial' alluvia.
7. Boulder-clays and terminal moraines of mountain regions; 100 feet beach of Scotland; Arctic plant-beds.
8. Lower buried forest.
9. Peat overlying 'lower buried forest;' Carse-clays and raised beaches; valley-moraines; corrie-moraines.
10. Upper buried forest.
11. Peat overlying 'upper buried forest;' low-level raised beaches; high-level valley-moraines and corrie-moraines.

1. Marine deposits with pronounced arctic fauna.
2. Temperate flora. *Elephas meridionalis*, *E. antiquus*, *Rhinoceros etruscus*, hippopotamus, &c.
3. Ground-moraines, &c., of most extensive ice sheet.
4. Northern and temperate flora and fauna; *Elephas primigenius*, *Rhinoceros tichorhinus*; reindeer and hippopotamus. *Elephas antiquus*, *Rhinoceros leptorhinus*. Irish deer, grizzly bear, lion, hyæna, &c.
5. Ground-moraine of ice sheet which extended to the Midlands of England.
6. Temperate flora and fauna; Irish deer, red deer, *Bos primigenius*.
7. Moraine accumulations of district and large valley-glaciers; Arctic marine fauna; snow-line at 1,000 to 1,600 feet; Arctic flora.
8. Temperate flora and fauna.
9. Small glaciers in mountain regions; snow-line at 2,400 to 2,500 feet.
10. Temperate flora and fauna.
11. Small glaciers in the most elevated regions; snow-line at 3,500 feet.

---

[1] *The Great Ice Age* (1894), p. 613.

Professor J. Geikie, in his well-known book,[1] gives in all seriousness a most elaborate but impossible scheme of six Glacial periods and five mild inter-Glacial periods! These ingenious theories may be all very well for the student of geology to play with in his mind, but unfortunately they do a great deal of harm when put forward by distinguished geologists as if they were proved ; for those who, having no time for such study, look to a leading geologist for authority, naturally believe them.

Glacial Geology as here presented may seem to some readers rather a dry subject, but the best way to get rid of that notion is to go out among the mountains, or in the fields and highways, and examine glacial deposits for ourselves, to pick up the pebbles and see if we can tell where they came from; to notice the 'striated blocks,' telling plainly of glacier action ; to examine the sections of boulder clay—those who do this find a new world of interest opened out, and, like Charles Kingsley, can

> See in every hedge-row
> Marks of angels' feet ;
> Epics in each pebble
> Underneath our feet.

---

[1] *The Great Ice Age* (1894), p. 607.

## CHAPTER VI

### THE ANTIQUITY OF MAN

The proper study of mankind is man.—POPE.

ALL who accept the truth of the saying quoted above—and there are few who will not—must agree that the problem of the antiquity of the human race is one of the most interesting questions to which the mind of man can be applied. The elements for its final solution are not all known at present; and it will be necessary to wait for further discoveries before the answer can be given. But, for all that, those who have not yet studied the subject will be glad to hear what conclusions, if any, have been reached, and to know what is the evidence at present available. Now, in the first place, it is important to draw a sharp distinction between what is *possible*, or probable, and what is actually known. Geologists and palæontologists think it very probable that man originated at a somewhat early period in the great Tertiary era of the world's history. This idea is based upon the knowledge which has already been obtained of the great development of mammalian life which began in the Eocene, and, going on steadily in the succeeding Miocene and Pliocene epochs, attained its highest point in the Pleistocene.

The rise and progress of the apes in Miocene and Pliocene times gives us a kind of backward limit, for it is clear that to place the origin of man *before* that of the higher apes would be like placing the cart before the horse. If, as Darwin believes, the human race sprang from anthropomorphic apes, we are compelled to believe that we came after them. Reasoning of this kind points to the belief that the great epoch-making era in the history of our planet must not be placed further back in time than the Miocene period. It *may* have been later, but, unfortunately, our 'missing link' or ape-like ancestor has apparently not yet turned up—or rather has not yet *been* turned up—*i.e.* dug out of his unknown hiding-place in the strata. We must confine our attention to what is actually known, that is, to facts rather than fancies. (See Chap. VII.)

In the present chapter, therefore, we shall endeavour to give a brief account of the evidence at present collected, and to show in what direction it points, only premising, as before hinted, that geology has, at present, no final answer to give to the question. It will be better, in this case, to go backwards, and we must feel our way very cautiously as we approach those dim and distant times which preceded the Glacial period.

The subject naturally divides itself into two parts : (1) the antiquity of civilised man ; (2) that of uncivilised man. The distinction may at first sight appear rather arbitrary, but it may be said of civilised man that he was capable of founding cities and making laws. Such a distinction was familiar to Horace, as the passage quoted on page 1 will show. It is known that the Great Pyramid of Egypt is nearly 6,000 years old according to the most recent researches. But there are others, like that of Sakkara, which may well claim to be slightly older.[1] It is asserted by modern Egyptologists that civilisation in the delta of the Nile goes back some 10,000 years !

Mr. William Peck, in his 'Handbook and Atlas of Astronomy,' has put forward a theory with regard to the origin of the signs of the Zodiac, which, if true, would give them an antiquity of over 14,000 years ! He maintains that they come from Egypt, and that, when first invented, each of the twelve zodiacal constellations had a direct reference to the phenomena of the seasons and the rise and fall of the Nile, which, together with the

[1] See Maspero's *Dawn of Civilisation.*

sun, was, to the ancient inhabitants of the delta, a deity, and therefore an object of worship. At the present time no such connection can be traced. The signs have lost their meaning. We shall eagerly look for Mr. Peck's promised work on the subject; and, *if* he can succeed in establishing his theory, he will do a great service to archæology by giving a clue to the date of the earliest civilised community, for we take it that any community which observed the stars, gave them names, and recorded their positions with regard to the sun, may claim to be called a civilised people. The lower part of the Euphrates Valley, inhabited first by Akkadians and then by the Babylonians, occupies a similar latitude to that of Egypt; and, since there was from early days a good deal of intercourse between the two peoples, it would not affect the theory if it should be proved that the signs of the Zodiac originated among the people of the valley of the Euphrates.[1]

No doubt to all early races, as to savages now, the moon was the first time-keeper. Day and night are such obvious phenomena, and so closely connected with the life of man, that we need scarcely speak of the day as a division of time. The month would be the next one. It would not take long to discover that in a period of twenty-eight days the moon goes through all her phases. Then, by noticing her position among the stars from day to day, as seen in her rising and setting, it would be discovered that after twenty-eight days she came back to the same group of stars. The apparent path of the sun among the stars is not so easily observed, and so this discovery came later. But by observing the stars near the solar orb at sunrise and sunset, people would find out that it too had a track of its own, and that in the course of the four seasons of spring, summer, autumn, and winter this great light-giving orb returned to the same group, only to begin over again its journey among the stars. This gives the year;

---

[1] Both Professor Flinders Petrie and Sir Henry H. Howorth say they reject the theory. The present writer does not wish to bind himself to it, but the subject is interesting, and therefore may be introduced here.

and it would easily be noticed that the time occupied was thirteen lunar months, each of twenty-eight days (*i.e.* 364 days). Later on the month of thirty days was obtained from a combination of the apparent motion of the two orbs.

The sun's track would now be divided into twelve spaces instead of thirteen, and these, of course, would be marked out by certain groups of stars or constellations, the names of which, as we shall see, give most interesting evidence with regard to the origin of the signs of the Zodiac, their connection with the seasons, and the occupations which respectively belong to them. Now, as the product of thirty and twelve is 360, the year formerly was one of 360 days. Hence the division of the circle described by the sun, and so of all circles, into 360 degrees. And it appears that the constellations of the Zodiac were in all probability the first invented, these being simply at first a rude kind of calendar dividing the year into twelve parts or months, and by the nature of the constellation figure indicating, at the time when the sun was in its midst, what the particular month was noted for.

A celestial sphere of the heavens made in marble and resting on the figure of Atlas has been discovered at Rome, which is believed to be an exact copy of the celebrated sphere of Eudoxus. The various constellations are clearly depicted on it, and, as Eudoxus travelled to Egypt about the year 380 B.C., it appears probable that the Greeks obtained the names of the constellations from the Egyptians, from whom they learned so much in other ways. This writer was the author of two important works, showing the connection between the weather and the different positions of the constellations in the sky with regard to the sun. Every one who is in the habit of watching the stars must have noticed that at any given time, say eight o'clock in the evening, we notice various constellations rising or setting in the sky. And we connect these with the seasons, because we see some at this time in the winter months, while a different set are to be seen in the summer months. This is due to the earth's revolution round the sun every year.

But the ancients thought it was the sun that moved and not the earth. The track along which the sun *appears to move* is known as the ecliptic. It is really the plane in which the earth performs her revolution round the sun, or, in other words, an imaginary plane cutting through the centres of the earth and of the sun. The writings of Eudoxus are lost, but all that is known of them is gathered from the poem of Aratos, who wrote about a century later. This poem seems to contain a versification of the descriptions of the constellations given by Eudoxus in the 'Phainomena.' St. Paul quoted from the poem, which is one of the most popular of ancient works.

The precession of the Equinoxes is a very slow change brought about by a gyratory movement of the whole earth, something like the wobbling of a top before it ceases to spin. The whole axis describes a cone in space, and the North Pole therefore describes a circle. Picture that pole produced for millions of miles into space, and you can see that the end of it would describe a circle among the stars. Now that circle is described in 26,000 years. The period spoken of in our last chapter, viz. 21,000 years, is due to precession combined with another movement known as the 'revolution of the apsides.' To us on the earth this produces a slow apparent movement of the stars, and a displacement of the constellation figures in the heavens. Many of these imaginary figures, therefore, which, at the time of their invention, appeared in their natural postures are now much altered in position, and in some cases completely inverted. In the same way the pole star keeps changing, and the one which is now the pole star was not the pole star (*i.e.* near, or at the pole of the heavens) at the time the Great Pyramid was built. According to Mr. Peck, the constellations of Pegasus, Andromeda, and Cassiopeia are inverted. 'We have shown,' he says, 'that the Zodiac must originally have been employed as a division of the solar year into twelve parts, the individual constellations representing, in all probability, the season of the year, the nature of the prevailing weather, or the principal occupation at the time when each group was in conjunction with the sun, or when the solar orb was passing amongst its stars. For the constellation figures do not seem to have been suggested by a resemblance of the various star groups to familiar objects, because in most cases they have no resemblance whatever to the object named, but each of the old groups was identified with objects typical of what was continually taking place either in the heavens or on the earth, in particular with the struggles, victories, and defeats of the "Great Sun God," in his encounter with the evil powers of darkness, while performing his annual journey round the star sphere. If this, then, be the object for which the Zodiac was invented, one would expect that its constellation figures would be strikingly representative of some particular country, and, if of Egypt, they would be intimately connected with the Nile ; for from the remotest antiquity Egypt has been considered as solely dependent on that great river, as, from the regular overflow of its water, this country is in reality " The Gift of the Nile."'

It would, therefore, be naturally expected that, if the signs of the Zodiac originated in Egypt, many of the constellations therein would be connected with the rise and fall of this river, as it takes place regularly each year ; and that such is the case we shall be able to show. But, first, it will be well to point out what are the main features of the Egyptian seasons, at what times in the year they occur, and how they are connected with the state of the Nile, or with the weather, as the case may be ; otherwise it would be impossible to understand the very striking significance of the various signs of the Zodiac, and to see how admirably they seem to have been chosen to mark all the important natural events in the cycle of the year.

In Egypt, the ancient solar year appears to have begun at the lower or winter solstice, which was immediately followed by the sowing of seed. Three months later, at our spring (vernal equinox) in March, the harvest took place. This seems early; but we must remember that the soil of Egypt is very fertile, owing to the rich deposit left by the Nile in flood, and that the sun has greater power there, in latitude 30° N., than in Northern Europe. After the harvest, the people were occupied in weighing and measuring their crops, and paying tithes and taxes to Pharaoh's officers, the king being the landlord of the whole country. For the next three months things were very unpleasant, on account of the deadly south wind, or *Khamsin*, which still blows for fifty days. This was the time when plagues occurred. By the middle of June (summer solstice) the elements were more favourable, and the Nile, in ordinary seasons, began to rise, continuing to do so until the whole land was flooded in August. The limit of the flood was reached about the end of September, and it was not until October that the country once more resumed its usual appearance, and the great river was confined to its ordinary channel. In November, when a rich alluvial sediment covered the land, cultivation began, followed by the sowing of seed in December.

At the present time, when the sun is among any of these groups—*i.e.* in conjunction with them—events are taking place in Egypt quite different from those symbolised by the groups themselves; and it is only by working backwards to 14,500 years ago that we get the desired correspondence. At that distant period the bright star Vega, in the constellation Lyra, was then very near the celestial pole, and another bright star, Spica, in *Virgo*, was near the vernal equinoxial point. This may appear at first sight rather a startling conclusion; but in no other way, apparently, can we get the required correspondence between the sun's monthly position among the stars and the state of affairs in the country of Egypt—*i.e.* harvest weather, &c.

Let us begin with the constellation Virgo. At the time we are now speaking of, it was harvest time when the sun was among the stars forming that group, and, as already stated, harvest took place

# THE ANTIQUITY OF MAN

at the vernal equinox, when the sun appears to cross the celestial equator.[1]

On inquiring into the significance of this constellation, we find that the Virgin has long been considered to be a purely harvest constellation. Arato says, in his poem, that she 'carries in her hand the brilliant ear of corn '—*i.e.* the bright star Spica. Next comes *Libra*, the scales, if we go in the order in which the sun goes through the Zodiacal constellations in his annual journey. This is not, as some appear to think, a more recently named constellation. Coming after 'the gleaning maid,' it is clearly symbolical of weighing the crops, balancing accounts, and paying taxes. Be it remembered that in ancient Egypt there was no coinage, but gold and silver were weighed out, as Abraham weighed out money to the Hittites when he bought the cave of Machpelah.

The sun next entered the group *Scorpio* and afterwards *Sagittarius*. These symbols are symbolical of highly important natural events which took place every year during the two months preceding the summer solstice or midsummer. For it was during April that the solar orb began a severe conflict with the fatal south wind, which brought with it death and destruction. At times, the disc would be nearly obscured by the large amount of blown sand from the desert. Such an event would be well symbolised by the sun's entry into the stars forming the claws of the Scorpion, fitting symbol of destruction and evil. This deadly struggle went on until the month of May, and not till then did the sun triumph over his enemy. And how could such a triumph be better symbolised than by Sagittarius, the archer, shooting his fatal arrow at the Scorpion? He, again, is the solar hero victorious

---

[1] For the convenience of teachers and others, we would like to add that Mr. George Philip, in Fleet Street, has published a very excellent model of the earth, surrounded by a glass globe, on which the celestial equator, the ecliptic, and the signs of the Zodiac are all clearly marked. It is important to point out to young people that this yearly motion of the sun is only an apparent movement caused by the earth's yearly revolution round the sun.

in his struggle with evil, an archetype of our St. George and the dragon.

In some cases, the cuneiform and even the more primitive Akkadian words for the constellations of the Zodiac show corresponding meanings. Thus, in the cuneiform, Sagittarius is called 'the strong one,' and 'the illuminator of the great city;' Capricornus, the goat, was entered in June, the month of the summer solstice. Professor Sayce says that the Euphratean name for this month is 'Father of Light,' and this is to be explained by the fact that during that month the days were longest and illuminated to the fullest extent. But in the time of Eudoxus this same constellation occupied a position nearly opposite, near the winter solstice, 'where retrogrades the solar night,' as the poem of Aratos says. And it does so still; for, on referring to 'Whitaker's Almanack,' the reader will see that the sun enters in the latter part of December. Now, as goats are in the habit of climbing, and live among mountains, the constellation Capricornus appears to be connected with the fact that the sun had climbed up to the highest point in its yearly journey in the sky. An interesting point about the goat is that it once had a fish's tail, as on a Babylonian stone, and also on a MS. in the British Museum. The sun entered the tail just when the Nile began to rise.

The cause of the rise and fall of the river was a great mystery to the ancients. That loquacious old gentleman, Herodotus, speaks of it, and mentions some of the suggestions he had heard. He rejects the true explanation, namely, that it is due to the melting of snow on certain mountains far away south, which fed the great river with their streams and snows. The Nile went on rising in July, and during that month the sun entered Aquarius, the water god, and at that time its waters were being literally 'poured out over the land.' Other constellations near testify to the same event. The southern fish is swimming in the stream, and the dreaded sea-monster (*cetus*) playing in the waters.

The next Zodiacal constellation is the fishes (*Pisces*), and they are very appropriate, considering that fish would then be plentiful. The autumnal equinox also took place when the sun was in this constellation, and that, of course, would be September. The Nile waters now began to retire, and in less than a month the sun had passed through *Aries*—the ram and sheep could be safely sent out to pasture. The long winding constellation Eridanus, the river, would be near, and probably it also represents the water god, if not the Nile itself, and by this time the river would be once more confined to its proper channel.

At this season a ram was probably sacrificed, and the constellation may be a reminder of this. It was the month for cultivating the land, and, as the ploughing was done by oxen, the bull would be an appropriate symbol for expressing the occupation of the people at that time of the year. The Pleiades, which are in this same group, appear always to have been connected with agriculture, as is shown by the poem of Aratos. *Gemini*, the twins, was the next constellation traversed by the sun; it was known to the Greeks as Castor and Pollux; but to the Egyptians, simply as a pair of kids. Here we seem to have a reference to the lambing season, which was in November. *Auriga*, a neighbouring constellation not in the Zodiacal circle, is a figure of a man with two kids in his arms, and so appears to be a repetition of the same idea.

The sun now began the lowest, and therefore the most critical part of its whole yearly journey. Hitherto, it has been supposed that when *Cancer* received its name it occupied the place of the higher or summer solstice, and, as the creature walks backwards, we should rather say sidewards, it might be taken to be symbolical of retrogression on the part of the sun. But Mr. Peck shows that such is not the case, and that it occupied precisely the opposite position, namely, that of the winter solstice, when the days were shortest, and a conflict seemed to be taking place between the rival powers of light and darkness. He thinks that the crab itself is really an Egyptian scarabæus; if so, we have at once a fitting symbol of resurrection and triumph, because, with the ancient Egyptians that insect was always a token of self-creation, and so of continuance of life. To them it was very sacred, and closely connected with the firm belief which they had in the immortality of the human soul. *Leo*, the lion, was the next constellation entered, and the entrance would take place in January, and, as Leo is a large constellation, it would take the sun a long time to get through it. Probably, then, the solar orb would not be leaving it till near the vernal equinox, two months later. Towards the middle of February the sun would be getting more and more powerful, and consequently the crops would be growing fast, or, in other words, the solar hero would be moving in strength like a lion. The passing of the vernal equinox would mark the sun's triumph, and a lion would be a good symbol for victory. On crossing this point, the sun passed into that upper half of the celestial sphere which was one of light and life.

We have now followed the sun all along its annual apparent path among the stars, and we recommend those of our readers who are interested in this line of argument to consult Mr. Peck's useful and copiously illustrated book. If the reasoning is sound, the theory offers a better explanation of the signs of the Zodiac than has yet been found, and, at the same time, gives a basis for calculating the age of the oldest civilisation in the world. Of the late Archbishop Ussher's chronology, which some people appear to think is sanctioned by the Scriptures, we have already spoken in the preface. No educated person in these days believes that the world was made 4,000 years ago.

With regard to the chronology of a Bronze Age, there is no doubt that the discovery of the art of making this alloy of copper and tin must be very ancient. But we have really no data on which to work out its antiquity. In all probability, according to one of the best authorities, it was in use at the time the Pyramids of Egypt were built, and that would make it some 6,000 years old. How long the Bronze Age lasted, and at what time it was replaced in Europe by iron, are questions which must be reserved for a later chapter.

We now pass on to consider various interesting and more or less converging lines of evidence with regard to the antiquity of man. Assuming, as we have done all through, in spite of a conflict of opinion among geologists, that Palæolithic man lived in the Glacial period, it remains to be considered whether there are any means of calculating, even in a rough manner, the number of years that have elapsed since the men of the Older Stone Age hunted the mammoth, reindeer, and other beasts that have either become extinct or have returned to more northerly regions. We have seen that there is no reliance to be placed on astronomical considerations, and calculations therefrom with regard to the remoteness of the period in question, but it does seem as if some help in the required direction may be obtained from purely geological considerations. In some cases it is even possible to obtain rough calculations of the length of time necessary for certain geological operations to have produced the effects which we know they have produced. These, it must be carefully pointed out, are not to be taken too seriously, much less to be relied upon as at all accurate, but only as affording certain *useful limits*, and thereby showing that the demands of certain geologists for 80,000 years or more are quite unjustifiable, either by reason or by the facts.

Among the various subjects about which calculations have been made, the following are perhaps the most important :

(1) The age of the delta of the Nile.
(2) The age of the delta of the Mississippi.
(3) The age of the Niagara Falls.
(4) The rate of excavation of river valleys.
(5) Pedestal boulders.
(6) The flow of Greenland ice.
(7) The formation of stalagmite in caves.
(8) The time required for the growth of peat-beds.

Some of these subjects have been very fully discussed in the works of Sir Charles Lyell, Sir John Lubbock ('Prehistoric Times') and Professor Boyd Dawkins ('Cave-hunting' and 'Early Man in Britain.') It will not be necessary, therefore, to discuss these calculations in detail, and it is quite sufficient for our present purpose if we only give the results which have been obtained, together with a general idea of the method pursued.

To begin with the case of the Nile delta (1). Mr. Horner made some researches (undertaken at the joint expense of the Royal Society and the Egyptian Government) in order to determine the rate at which the delta is being built up by the annual alluvial deposit of the river. The French *savants* accompanying Napoleon's expedition arrived at the conclusion that the average rate of growth was five inches in a century. Mr. Horner, however, preferred to make independent observations. Taking the obelisk at Heliopolis, he found that in 1850 eleven feet of alluvium had been deposited since it was placed there, about the year 2300 B.C. —*i.e.* in 4,150 years—which gives a rate of 3·18 inches per century. Applying a similar method to the statue of Rameses II. (the oppressor of the Jews), he found the present surface of the delta to be 10 ft. 6¾ in. above the base of the platform on which the statue stood. Assuming this platform to have been 14¾ in. below the surface of the ground at the time it was laid, he found that a deposit of 9 ft. 4 in. had been since formed. Rameses II. is supposed by Lepsius to have reigned from 1394 to 1328 B.C. Taking the middle of his reign (1361) as the time when the statue was set up, we obtain an antiquity (up to the year 1850 A.D.) of 3,211 years, and consequently a rate of 3½ inches per century.

Having thus obtained an average rate of growth for the whole deposit, Mr. Horner dug pits to a considerable depth. Pottery was found in one of these at the great depth of thirty-nine feet, and resting directly on desert sand. Hence, he obtained an antiquity of at least 13,000 years for the whole depth of alluvium, that is, for the delta itself. It is not to be denied that there are some elements of uncertainty in the calculation;

but if, out of a number of attempts of the kind, some amount of agreement results, it may justly be considered that the methods employed are more or less trustworthy.[1] The operations themselves are all such as a geologist may well believe to have been confined to the latest, or Pleistocene period, during which the hunters of the Early Stone Age lived. And so, if that conclusion is sound, it is clear that we have in all the above operations measures of the antiquity of the period in question. Some of them appear to have been obviously confined to a time later than the Glacial period; for instance, the pedestal boulders left by the ancient glaciers. Again, the peat beds may safely be assumed to be also post-glacial. Then again, with regard to the excavation of river-valleys, we know how much certain river-valleys have been deepened since glacial times, and the amount of geological work thus done *ought* to be a measure of time. Some of the operations, on the other hand, such as the cutting of the gorge of Niagara, may take us back right through the Glacial period, but our object is merely to obtain limits, and to show that the enormous demands of some geologists with regard to the antiquity of that period, and therefore of the human race, are not confirmed by calculations of this kind.

Passing on to the case of the Mississippi (2), attempts were made some years ago to calculate the *least* amount of time in which so vast an amount of solid matter could have been laid down as that which then constituted the delta. Messrs. Humphrey and Abbot made a careful survey, and estimated the area at 12,000 square miles, the average depth of the fluviatile deposits at 528 feet, or one-tenth of a mile, and the quantity of sediment in suspension as $\frac{1}{1245}$. Knowing approximately, from their own measurements, the annual discharge of water, and allowing liberally for the large amount of matter in the shape of sand and gravel pushed along the bed of the river, Lyell came to

[1] Sir Henry Howorth, on reading this, informs us that the pottery was Macedonian. If so, the calculation fails.

the conclusion that the delta must be at least as old as 33,500 years, and probably much more.

Attempts have been made to calculate the age of the famous Falls of Niagara (3). At first sight this problem does not appear to be in any way connected with the question of the antiquity of man. Geologists, however, believe that they have evidence to show that a great part, if not the whole, of the ravine in which the river flows has been excavated since the Glacial period. That fact, if established, at once brings it into line with the question we are considering. It therefore becomes interesting to see if there is any way of calculating, however roughly, the time required for such an operation. Now, the gorge of the Niagara extends for seven miles back from a line of cliff formed by the escarpment of the massive Niagara limestone; and there can be no doubt that the whole of this long ravine has been slowly cut out by the action of the river in wearing away the edge of the limestone rock over which it falls continually. Under the limestone are soft shales, which the water washes away much more easily than the hard limestone, and so the latter stands out and makes a fall. We do not know the circumstances which controlled the rate at which the Falls were cut back, but among these the most important are the volume of water, the hardness of the rock, and the lie, or 'dip,' of the rocks. The present position of the Falls seems to be rather favourable for rapid cutting away. Bakewell, from the testimony of old residents, and from historical notices, inferred the rate of recession of the Falls to be three feet a year. Lyell, from personal observation, came to the conclusion that it was more like one foot a year. Now, seven miles contain 36,960 feet, which is not far off from Lyell's estimate of 35,000 years.[1]

More recent estimates, on the other hand, come nearer to that of Bakewell—namely, three feet a year. On that basis we obtain

[1] Lyell, *Travels in North America*, i. 32, and ii. 93; also *Principles of Geology*, i. 358.

for the Falls an antiquity of about 12,000 years only. Some say the 'Horseshoe Falls' have receded 104 feet in forty-eight years, or just about two feet a year. On that basis the antiquity would be 18,480 years.[1] Whichever of these estimates be adopted, they are certainly useful in tending to confirm the views expressed in a former chapter with regard to the comparative nearness in time of the Glacial period, and therefore to discredit the views of Croll and J. Geikie, who believe that 80,000 years have elapsed since the end of the Glacial period!

[1] Sir Henry Howorth again comes to the rescue, and informs us that Mr. Spencer and other American geologists have re-examined the question with different results to those of Lyell.

## CHAPTER VII

THE ANTIQUITY OF MAN—(*continued*)

As Nature first made man
When wild in woods the noble savage ran.—DRYDEN.

WE must pass on now to the general question of the cutting down of river valleys (4). If only some rough estimate were possible of the average rate at which the process has been going on since man inhabited certain caves on river banks which are at considerable heights above the stream, we should be provided with a means of calculating the time that has elapsed since the Older Stone Age. Again, there are the high-level gravels, containing Palæolithic implements such as those of the Thames, to be seen at Hampstead Heath; and it would be very desirable to obtain some idea of the rate at which Father Thames, or the Somme, have been excavating their beds since then. Sir Joseph Prestwich says that such rivers as these do not seem to have cut down more than 80 to 100 feet in that time. The older school of geologists would have assumed, in accordance with strict ideas on the subject of uniformity of action by geological agents (as expounded by Lyell), that many thousands of years would be required for this cutting-down process. But those of the younger school, now arising, who adopt to some extent the views of Prestwich, see no difficulty in believing that 10,000 to 15,000 years might be quite sufficient. It may well be believed that the rainfall was considerably greater during the vast melting that took place at the end of the Glacial period; and it would be absurd to apply the *present rate* of 'denudation' to problems of this nature.

# THE ANTIQUITY OF MAN

Now, if a period of 80,000 years or more has elapsed since the high-level gravels began to be formed, how exceedingly slow must the rate of denudation have been! Let us see what this means. It means that in a great deal more than 80,000 years the rivers of the South and North of France have only succeeded in excavating their valleys to a depth of from 80 to 100 feet. Taking the lower limit of 80 feet, we arrive at the result that the process of cutting down or carving out their valleys has only been since going on at the absurdly low rate of one foot in 1,000 years! Now, we ask, is any geologist of the present day prepared to accept such a conclusion in the face of the fact—admitted even by the late Dr. Croll himself—that denudation took place very much faster for some time after the Glacial period? Dr. Croll once said: 'If the rate of denudation be at present so great, what must it have been during the Glacial period?'[1]

Perhaps we cannot better show the absurdity of this result than by contrasting it with the important and elaborate results arrived at with regard to the *general* lowering of the surface of a continent at the present day by the forces of denudation.[2]

The result of careful measurements and observations on the river Mississippi, with regard to the amount of transportation it accomplishes each year, appears to show that such an amount of solid rock and soil is being removed from the surface of the great area (7,459,267,200 square miles) drained by that mighty river as would be represented by a general lowering of it to the extent of the $\frac{1}{6000}$th of a foot, or one foot in 6,000 years. Other large rivers have been subjected to similar tests, but since the observations are more complete in the case of the Mississippi, and as that river drains so extensive an area, embracing so many varieties of

[1] *Climate and Time*, ch. xx.
[2] Some writers speak of a 'Pluvial period' following that of the glaciers, when the force of rivers was twenty times greater than it now is, and the rivers were twenty times as large. We have understated our own argument by assuming the post-Glacial date of the higher gravels, whereas we believe they were Glacial.

climate, rock and soil, the result is more likely to be trustworthy, and so geologists generally assume this as a fair average rate of denudation for the temperate region of the Northern Hemisphere. Now the denudation of a vast area like that is a very different thing to the carving out of a river-valley. The latter is a comparatively rapid action. And yet, according to the argument used above, we should be compelled to believe that the Thames, for instance, had in late geological times cut down its bed at a rate only six times greater than this slow process of denudation over a whole country—*i.e.* one foot in 1,000 years instead of one foot in 6,000 years. Such a conclusion seems to the present writer a veritable *reductio ad absurdum*, and, as far as he is aware, the argument has not been put in this way before.

We will now state the argument somewhat differently. Let us take the more reasonable limit of 15,000 years for the antiquity of Palæolithic man, instead of 80,000 years. Then, assuming the geological work of the rivers to have been done in that time, we obtain a much more satisfactory estimate of the rate at which valleys have been cut down. Thus 15 is to 80 nearly as 1 is to 5. If the rate of cutting down had been five times faster, it would be five feet in 1,000 years, instead of one foot in that time. This is only an average rate, but, even as such, it seems a slow one. But we might be justified in putting the case much more strongly if we were to take the age of the gravels in question as glacial. In that case we might take quite double the above estimate of their antiquity, according to Croll—*i.e.* 180,000 years—since he considered that the Glacial period began 250,000 years ago. Now, if the Thames and Somme have only lowered their valleys 80 feet in 250,000 years, the rate of cutting down must have been only one-third of a foot in 1,000 years! It works out thus:

$$\frac{80 \times 1,000}{250,000} = \text{about } \tfrac{1}{3}.$$

With regard to the question of the geological age of the high-level gravels, Sir Joseph Prestwich says : [1] ' My original impression with respect to the valley of the Somme was that the high-level gravels originated in later glacial times ; that the intermediate stages and terraces were formed during the excavation of the valley as a resultant of the late-glacial and post-glacial floods ; and that the low-level gravels formed the concluding stage of these conditions. But since then the whole series has generally been considered as post-glacial. The older date would also better agree with the classification, based mainly on archæological considerations of M. de Mortillet, though I should not place, as he does, these beds at the beginning of the Quaternary period. Possibly, however, our base line differs.' And, after alluding to certain discoveries, he continues : ' I am therefore led to conclude that the high-level beds of the Somme Valley at Amiens, of the Seine in the neighbourhood of Paris, and of the Avon at Salisbury, together with certain caves, date back to Glacial times, though not to its earliest stages.' [2]

Before leaving the subject of denudation, it may be worth while to consider the phenomenon of certain boulders standing on a pedestal of rock, and known as 'pedestal boulders' (5). These are to be seen in the Lake District and in parts of Yorkshire, resting on the formation known as mountain limestone. Evidently they are 'erratics,' like those mentioned in Chapter IV., and have travelled some distance from the parent rock of which they once formed a part. But it is not a very easy matter to say how they came to rest on the spots they now occupy. Some geologists suggested that they had been brought there by icebergs and left stranded. But this theory necessitated a belief in the submergence of the district during the Glacial period—a theory which presents many difficulties. Others suggested that the pedestal

[1] *Controverted Questions of Geology*, p. 44.
[2] Some geologists accept Mr. Skertchley's supposed discovery of Palæolithic flint implements under boulder clay in East Anglia. But they were *not* associated with extinct animals, and the bones found with them belonged to recent species. We cannot therefore accept his conclusion (see Chapter IV.).

boulders had been washed out of boulder clay, of which all other trace had disappeared. This also involves no small difficulties.

The subject has been investigated by our friend, Professor T. McKenny Hughes, from whom we learned much at Cambridge. His explanation is far more satisfactory. In a paper read before the Geological Society,[1] he showed that they were gently laid to rest where we now find them by the recession of the ice sheet that once covered that part of the country, or, in other words, by the melting of the glaciers. In some cases they appear to have been carried up and over the brow of a hill. A figure of one of these blocks is given in Prestwich's 'Geology,' vol. i. p. 45. It represents one of the numerous boulders of Silurian slates strewn on the limestone hills round Clapham and Settle, in Yorkshire. In this case, as in others, the glaciated limestone, where not protected by blocks, having been slowly dissolved away by the action of rain water, has left the blocks perched on pedestals from one to three feet high. The present writer was much struck with this figure, and it suggested to him the idea that we have here a chronometer wherewith to gauge the length of time that has elapsed since the retreat of the glaciers.

This is the idea which Professor Prestwich works out in his paper.[2] He says 'the tops of the pedestals are smoothed and striated by glacier action.' 'Thus we have the print of the old glacier stereotyped as it were.' Generally the pedestals are twelve to twenty inches high; sometimes only three to seven inches. Their average height was from one to two feet. In the discussion which followed this paper, Professor Hughes said he thought there was too much uncertainty in the rate of erosion to allow of calculations such as might prove of any value. But, at all events, we think it showed that Sir Joseph Prestwich was right in believing that we have here sufficient evidence to entirely upset the astronomical calculations so much in vogue with certain geologists, and

[1] *Quart. Journ. Geol. Soc.*, xlii. p. 527.
[2] *Controverted Questions in Geology*, p. 44.

to prove that nothing like 80,000 years can possibly have elapsed since the old glaciers melted away. Again, we would ask, as we did before when speaking of the deepening of the old river-valleys, can any geologist in his senses believe that in 80,000 years the platform of limestone on which these boulders rest would not have been lowered by atmospheric denudation to a much greater extent than from one to two feet?

Another strong argument brought forward by Prestwich and others is based on the same idea. They appeal to the wonderful way in which the glacial striæ have been preserved. How could we possibly expect to find any trace of these delicate surface-markings if they had been exposed to wind and rain, storm and sunshine, for anything like the length of time just mentioned? This argument appears to the writer quite unanswerable, and, as far as he knows, no geologist has even attempted to answer it !

The flow of Greenland ice (6) is a subject from which the same distinguished geologist has attempted to obtain materials for calculating the length of time which was required for the advance of the ice sheet over a large part of Britain, say 500 miles.[1] Recent observations on the Greenland glaciers appear to show that the continental ice advances at the rate of one mile in eight or twelve years; so that the flow of the old British glaciers, or so-called ice sheet, might have taken place in 4,000 to 6,000 years.

Calculations based upon the advance of Alpine glaciers at the present time are of no use whatever, for they move much more slowly. According to the conclusions of Prestwich, the whole length of time required for the formation and extension of the great ice sheet of Europe and North America need not be extended beyond 25,000 years at the utmost, including in this estimate the time during which the cold was increasing and that during which it was getting less.

---

[1] According to Bonney, Howorth, and other geologists, a flow of 500 miles is an impossibility. The writer agrees, but that does not alter the argument—merely as an argument.

In a paper by Mr. C. J. Russell some important data are furnished with regard to the rate at which the continental ice retreats. It appears that in Alaska 'the glaciers are slowly retreating, and probably have been retreating for 100 or 150 years. The amount of this recession in the case of the glaciers at the head of Yakutat Bay is known to be four or five miles, and at the head of Glacier Bay the retreat is thought to have been not less than fifteen miles during the past century.' Applying this rate of retreat to the old British glaciers, and computing the area at 800 × 800 miles, Prestwich arrives at the conclusion that it could have entirely disappeared within the limits of 3,000 to 8,000 years. These calculations are not brought forward as anything more than rough approximations; but they are certainly useful as showing that the old estimates of 200,000 years or so can no longer be justified.

The final melting of the ice sheet, according to Prestwich, may have taken place as recently as 10,000 years ago. Mr. Mellard Reade has arrived at conclusions less in accordance with the new view, and calculates that a period of 57,000 years has elapsed since the deposition of the low-level boulder-clay of Lancashire, which he makes up as follows: 40,000 years for the elevation and denudation of the boulder-clay, 15,000 years for the formation of certain post-glacial beds, such as estuarine silt, peat, and forest beds (Cromer), and 2,500 years for the blown sand which in many places covers these deposits. We should like to know his answer to the question about the pedestal-boulders!

Some geologists have attempted to make calculations with regard to the rate at which deposits of stalagmite (7) may have grown in caves. But here again we must point out that any such attempts, if based on the rate of growth in recent times, must be quite useless, because it may be taken for granted that in the early part of the Stone Age, at least, the rainfall was very much greater than it is now. Consequently, the streams that flowed underground in the limestone districts, where alone caverns are found, were much more swollen, so that more carbonate of lime was deposited. Nor have we any reliable evidence with regard to temperature and other conditions, all of which are important elements in the problem. The subject has been fully discussed by Professor Boyd Dawkins,[1] who quotes observations that have been made to show that as much as a quarter of an inch may be deposited annually. At this rate, a layer of stalagmite twenty feet thick *might* be formed in 1,000 years.

And, lastly, calculations have been made with regard to the length of time required for the growth of peat-beds (8). For this purpose various observations on the present growth of peat are appealed to. Thus it has been found that the rate may be, in some cases, not more than one foot in a century; but it is commonly more, being sometimes as much as four, five, or even ten feet in that time. The local conditions of moisture and temperature vary so greatly as to show that we cannot trust those calculations which take for their basis a slow rate of growth in thickness.

Prestwich has also written on this subject, and discusses the time required for a Swiss deposit known as the Duernten beds. These are certainly 'inter-glacial' in one sense, for they are found truly intercalated between two glacial deposits.[2] They contain remains of trees, such as the pine, yew, birch, oak, and these are said to be of the same species as those now living. The animal remains include those of the *Elephas antiquus*, cave-bear, stag, and great urus, or *Bos primigenius*. The true reading of the story recorded here seems to be that a temporary retreat of a glacier occurred, such a retreat as that recorded of many Swiss glaciers; after which the above plants and animals flourished.

---

[1] *Cave-hunting*, p. 39.
[2] Sir Henry Howorth writes: 'This is quite exploded now. The Duernten beds are most distinctly pre-Glacial, that is, they underlie the so-called Glacial beds.'

Then the glacier advanced again, and so the remains got preserved between two glacier-moraines. But many geologists explain the facts differently, and prefer to believe that we have here evidence of one of those 'interglacial periods' which, as far as we can see, exist only in their own imagination. However, the question we have to consider here is, how long such a deposit may have been in forming. The bed varies from five to ten feet in thickness, and is rarely twelve feet. It should properly be called a lignite deposit, as there is more of wood than of peat. But it has been assumed that it required sixty feet of peaty matter to condense into twelve feet of lignite, or that it required five feet of peat to form one foot of lignite. Taking the rate of peat formation at one foot in a century (which, as above stated, is far too slow a rate), some geologists have arrived at the conclusion that it took 6,000 years to form the sixty feet of peat which they suppose was afterwards compressed into twelve feet of lignite. Prestwich, on the other hand, takes the original thickness of the peat at twenty-four feet and the probable rate of growth at two to four feet per century, and arrives at the conclusion that 600 to 1,000 years would be quite sufficient time for the growth of the Duernten beds—a conclusion which will appear to many geologists as much more reasonable.

Our object in dwelling on the above calculations has been to show, first, what has been done in this direction ; and, secondly, that in many cases the large demands for time which some geologists have made can be very greatly reduced. Lyell's theory of uniformity in geological action must not be followed too rigidly, but it ought to be allowed that in the days of Early Man the rate of change must have been considerably greater. In the opinion of the writer the preservation of glacial striæ is by itself quite sufficient to dispose of all such demands for many thousands of years since glacial times—a conclusion amply corroborated by the striking evidence of the 'pedestal-boulders.'

The chief value of the above calculations may be said to be

cumulative or collective. Any one of them by itself is not of very much value; but the reader will perceive that, taken together, they have a considerable value, because they all point in one direction, and that direction is one at variance with the extreme dates of the astronomical school of geologists, if we may so call them.

One more argument, which ought to carry a good deal of weight, may be mentioned here, and that is one which the naturalist and biologist will appreciate. It is this : that if anything like 80,000 to 100,000 years have elapsed since the Glacial period, we should expect to find much greater changes in the fauna and flora, and in the human race itself, than those of which there is evidence. We all, of course, believe that evolution works slowly, as all, or nearly all, great natural operations do, but we can hardly believe it works so slowly as all that !

We must now turn to quite another class of evidence, most of which is purely geological, though not all. In various places flint weapons have been reported as having been discovered in strata of Tertiary age, some of Miocene age, others of Pliocene age. And, in certain cases, fossil bones, supposed to be Pliocene, have been discovered, the surfaces of which were marked or grooved in such a way as to lead some authorities to suppose that they had been actually cut by human hands with flint implements ! Then there are some very ancient human skulls which have been the subject of much debate, about which we must say a few words. And, lastly, there is a very interesting recent discovery in Java.

Geologists have long been on the look-out for evidence of the existence of man, *homo sapiens*, in Tertiary times. The Pleistocene period is generally placed in a separate era altogether, that of the Quaternary. Of man's existence, then, there is plenty of evidence, as the reader will have learned from the accounts in previous chapters on the river-gravels, caves, &c. But the question naturally arises, whether man's existence can be traced further back. Did he live in the Tertiary era? That era is divided by geologists into four periods, as follows :

| | |
|---|---|
| Quaternary . . . . | Pleistocene |
| Tertiary . . . . . | Pliocene<br>Miocene<br>Oligocene<br>Eocene |

Thirty-one years ago Dr. Keith Falconer, a distinguished palæontologist, ventured to express the opinion that man may have existed at the latter end of the Tertiary period in Europe, but that we must look for his earlier, if not his earliest, appearance in India as far back as the year 1844. He was inclined to believe that the Indian mythology might furnish a clue in the direction required, and that some of the giant beasts mentioned therein might perhaps be, as it were, faint echoes through many centuries of the gigantic tortoises, pythons, and cranes, whose remains have been discovered sealed up within the Pliocene strata of the Siwalik hills of Northern India.[1]

This was only a speculation, and remains to this day 'not proven.' But probably it has been the means of stimulating search in that direction.

A year or two ago, Dr. Noetling, when mapping the Yenang-young Oilfield in Burmah, was interested in collecting remains of vertebrate animals, particularly in a conglomerate upwards of ten feet thick, which ran as a dull red band across the ravines and hills. As seen by him it appeared to underlie some 4,000 or more feet of Pliocene strata containing numerous fossil bones of the hipparion and other mammals belonging to that period. Here he found some chipped flakes partly projecting from the conglomerate. There was no doubt of their having been arti-

[1] The writer has dealt with this matter in his former work, *Extinct Monsters*, new ed., p. 172.

ficially chipped, and so evidently they were the work of human beings.[1]

Geologists were, of course, greatly interested in this announcement, for it seemed like the fulfilment of hopes long cherished. But, unfortunately, there was one little flaw in the evidence which entirely deprived the discovery of its supposed value. That flaw was detected by a very able geologist, our old schoolfellow Mr. R. D. Oldham, of the Indian Geological Survey, which his father directed so successfully for many years.[2]

Mr. Oldham,[3] in a communication on the subject, says: 'There are two distinct issues which must both be decided affirmatively before we can say that the existence of man in Miocene times in the Irawadi Valley has been proved. First, are the flakes of Miocene age? Secondly, are they of human origin? The second I do not propose to discuss; but, with regard to the first, the statement that a flake was found partially embedded in the rock requires, in the circumstances of the present case, an explanation. The site is on a spur running out into one of the valleys which have been cut back into the plateau; the crest of the spur falls somewhat rapidly and then rises slightly to the outcrop of the ferruginous conglomerate, whose exposure on the crest of the spur is, to the best of my recollection, about fifty feet long

---

[1] See *Natural Science*, September 1894, p. 345.

[2] In justice to a very singularly gifted teacher, we may perhaps be allowed to record here a little personal reminiscence. The present writer and Mr. Oldham both learned geology at Rugby from Mr. James M. Wilson, and often sat side by side at his lectures on that science, delivered in a manner which stimulated all who had any taste for natural science or any love of Nature. To him we brought all our difficulties, as well as the fossils we collected from pits in the neighbourhood, and we were always sure of guidance and warm encouragement. There is now a good collection of local fossils in the new Rugby School Museum, mostly collected by boys on half-holidays. Mr. Wilson afterwards became head master at Clifton College, and is now Archdeacon of Manchester. Many Rugbeians look back with pleasure on the old days when they first learned from him how to decipher that great stony record in which the history of the earth is preserved.

[3] *Natural Science*, vii. 201 (1895).

by eight to ten wide. No vestige of soil or sand is here, all having been removed by rain and wind, but there is a thin coating of ferruginous gravel overlying the solid rock, and it was on this surface, as pointed out to me by Dr. Noetling, that the flakes were found. Ordinarily, there would be no hesitation in ascribing anything found in this layer of loose material to the underlying rock; but it is not the same thing as finding a flake or fossil embedded in a bare vertical exposure below the level to which the rock had been loosened by weathering.' And he concludes by pointing out that there is at least a probability of the flakes having been dropped on the spur or washed down from the plateau above and subsequently becoming embedded in the weathered surface. This suspicion is confirmed by the fact that the implements in question are not confined to the outcrop of the conglomerate, but are scattered over the surface of the plateau above! And so, for the present, the discovery of Miocene man in Burmah remains but 'a consummation devoutly to be wished.'

Unfortunately, the same sad fate has befallen all the other supposed discoveries of Tertiary man which from time to time have made a flutter in the camp of geologists. Let us briefly review some of these cases. It is the more necessary to do so because certain writers to whom the public look for guidance in such matters would lead one to suppose that some at least of the cases are genuine.[1]

Take, for example, the discovery of the Abbé Bourgeois of worked flints apparently in Miocene strata at Thenay, in France, near Pontlery (Loir-et-Cher). The doubts in this case, as in the last, are of two kinds: (1) Are they true worked flints? (2) Do they come from Miocene strata? With regard to the first question there is less doubt, for some of them appear to be undoubted flint implements. The crux of the whole matter lies in the second question. It may easily happen that flint lying in a

---

[1] For example, Samuel Laing, *Human Origins,* p. 365; Edward Clodd, *The Story of Primitive Man,* p. 23.

superficial gravel may be displaced, and, falling into a pit, get mixed up with fossils from a lower stratum altogether, say of either Miocene or of Pliocene age. This is the simple explanation suggested by our greatest living authority, Sir Joseph Prestwich. And so the case is somewhat analogous to that of Burmah. For the present, then, it will be safer to suppose that the implements found by the Abbé Bourgeois were not *in situ*, but had fallen down from certain gravels at the top of the pit. The familiar expression, *Facilis descensus*, readily occurs, and in one sense the descent was both easy and natural. But in another sense it is rather 'a come-down,' to use a colloquial expression![1]

'At the Congress of archæologists and anthropologists held in Brussels in 1872,' says Professor James Geikie, 'opinion seemed to be equally divided for and against the human origin of the Miocene "flints" which M. l'Abbé Bourgeois submitted for examination. On the one side were MM. Worsaae, d'Omalius, Capellini, Mortillet, and other experts, who agreed with the Abbé; on the other side were MM. Steenstrup, Virchow, Fraas, and Desor, who opposed his views; while some, again, like M. Quatrefages, reserved their judgment, and were content to wait for additional evidence.' It is in some cases very difficult to detect the signs of human manufacture in rude implements; what one has to look for is evidence of design, or adherence to a certain type. Either kind of evidence is important, but the chief point is the evidence of design—*i.e.* of chips made with a certain object. But certain of the Thenay implements, to judge by the figures given of them, may certainly be considered as true implements. Some showed traces of the marks of fire.

We now pass on to the case of certain bones of *Elephas meridionalis* in sands or gravels at St.-Prest, near Chartres, in an old river-bed of Pliocene age. It is said that rude flint implements were found with the bones, such as might have been

[1] *Compte-rendu du Congrès International d'Anthropologie et d'Archéologie Préhistorique*, 1873, p. 81 (Brussels).

used by man to make the peculiar cuts or markings which were the subject of discussion. Some authorities were ready to believe that they really were made by human beings, and so thought that they had at last got on the track of Tertiary man. But Professor T. McKenny Hughes and Sir John Evans have both very wisely suggested that the marks in question might have been made by fishes' teeth![1] And it is much more probable that this is the true explanation. Looking at the figure given in Mr. S. Laing's 'Human Origins,' we are struck with the fact that the grooves seem nearly all to go in semicircular curves, instead of more or less straight, a fact which ought to tell against the idea that they were made by men. Sharks' teeth are very sharp, and so many of these supposed cut bones have been found that one is inclined to be rather suspicious. Professor Hughes gives a figure of a bone of a plesiosaurus so marked. Now if that had been a Pliocene instead of a secondary bone of the Kimmeridge Clay, probably some one would have been found to say that the marks were made by men. But in this case that is quite out of the question; and yet the marks are not at all unlike some of those which have been appealed to as evidence for the existence of man in the Tertiary period! Other cases have been much discussed; for example, a rhinoceros bone from the upper Pliocene strata of the Val d'Arno.

It only remains to say a few words about a recent discovery which has been much discussed in scientific journals, and even in newspapers, namely, the remains from Java, which some authorities think represent a veritable 'missing link' between man and the higher apes, such as anthropologists have been eagerly hoping to see for many a long day!

So far, nothing of that kind had been discovered. The skulls of Spy in Belgium (see p. 25), of Neanderthal, of Kanstadt, the fragment of a skull from Bury St. Edmunds, the jaws from Naulette, all point to a low type of humanity, but are nevertheless

[1] *Journ. Victoria Inst.*, May 6, 1889.

distinctly human (see Chapter III.). The above skulls show a low cranial arch, depressed frontal area, narrow foreheads, and immense ridges over the eyebrows. But take their capacity and one sees that they *must* be human. Besides, the Spy skeletons are complete, and it is easily seen that they are human, though rather bow-legged. That is a great help, because they assist us to a true interpretation of the other skulls, which are without the bones of the limbs. The cubic capacity (in centimetres) of the Neanderthal cranium is estimated at 1,200 c.c., whereas that of an anthropoid ape is only 500 c.c.

More than thirty years ago the whole question was most fully reviewed by the late Professor Huxley,[1] who showed that all fossil remains of man hitherto known were distinctly human in their characters and represented but a very slight approach to apes; while the oldest fossil remains of apes obtained from Tertiary strata were hardly nearer to man than the now existing chimpanzees, gorillas, or gibbons. A great gap clearly existed between apes and man.

Such was the state of affairs until 1894, when Dr. Eugene Dubois published his account of some few remains he had discovered in Java of a creature which he and others consider to be a more or less ape-like man (or woman).

The remains were discovered in the year 1891 by Dr. Dubois in certain strata on the south slope of the Kendeny hills in Java. These are thick layers of volcanic tuff consisting of clays, sands, and volcanic lapilli, cemented together and rearranged by river action; through these strata the river Bengawan has cut its channel. They are over 1,100 feet thick, and underneath them lie some marine deposits of Pliocene age. The true age of the strata appears to be still uncertain, although Dr. Dubois considers them to be Pliocene, as are those on which they rest. This is a very important point, and it is a pity

[1] Huxley, 'Man's Place in Nature,' *Nineteenth Century*, 1863, and *Collected Essays*, vol. vii.

that it has not yet been settled with certainty. But they are said to contain fossil bones of a gigantic pangolin, three times larger than the one now living in Java, of the stegodon, hippopotamus, hyæna, and several species of deer. Here it was that Dr. Dubois, while exploring for the Dutch Government, in September 1891 found the remains in question. First of all, he came across the cranium and one molar tooth. Next spring he found another molar tooth and a femur or thighbone in the same bed, but at some little distance off. That is a very important point in the matter, for it does not at all follow that the thigh-bone and the cranium were once possessed by the same individual, man or monkey, or whatever the creature was. The skull, if we judge by ordinary human proportions, was much too small for the big thigh-bone; but at the same time it is very much larger than the largest skull of a modern ape. Its capacity is estimated, for it is not complete, at two-thirds that of a well-formed European. The thigh-bone indicated considerable stature, and must have belonged to a creature that walked erect, as men do; the individual was estimated to have been five feet five inches in stature. Dr. Dubois read a paper on the subject before the Berlin Anthropological Society in January 1895,[1] and on that occasion Dr. Krause declared the tooth to be that of an ape, the skull that of a gibbon, and the femur human! The author of the paper concluded that the human and ape-like characters were so blended as to justify him in believing that he had discovered an erect ape-like man, which he thereupon christened the thing *Pithecanthropus erectus*. Virchow strongly opposed this view.

Dr. Cunningham, at a meeting of the Dublin Royal Society, described both the cranium and the femur as distinctly human. The skull, we ought to add, has a much depressed frontal region, and shows strong ridges over the eyebrows, as in the Neander-

---

[1] *Pithecanthropus erectus: eine menschenähnliche Uebergangsform aus Java*, by E. Dubois. (Batavia, 1894.)

thal specimen; but it also has more or less of a ridge down the centre, a feature which is much more strongly marked in the case of the apes. Skull capacity is an important test, and it may be mentioned here that the capacity of the skull of an average European is 1,400 to 1,500 c.c. As already stated, that of the highest apes is only 500 c.c., and that of the Neanderthal skull is estimated at 1,200 c.c. Now, that of the Java skull is estimated at 1,000 c.c., or 200 less than that of the Neanderthal skull. It is not, therefore, surprising that Dr. Dubois and others consider that it stands halfway between man and the apes, though somewhat nearer to man, as 1,000 is nearer to 1,200 than to 500. Hence the name 'ape-like man,' or pithecanthropus, does not appear to be altogether inappropriate.

The bones were exhibited at the International Zoological Congress at Leyden, together with a number of other bones and skulls for comparison. On this occasion the eminent American palæontologist Professor Marsh supported Dr. Dubois. In November 1895 the matter was discussed before the Royal Dublin Society,[1] and again before the Anthropological Institute of Great Britain and Ireland. The skull appears to be distinctly human, but, as Dr. Cunningham said, 'considerably lower than any human form at present known.' The anatomist Professor Rosenberg, however, considered the femur to be human, but saw in the skull that of a remarkably highly developed ape. The authorities are more agreed about the femur than about the skull, the former being distinctly human. Dr. Pearsal, a leading dental surgeon in Dublin, considered the teeth to be strikingly human, and yet larger than a man's, and their cusps are more developed, as in apes. Lastly, the thigh-bone tells a story of sickness and trouble to which the human race was ever prone. It shows indications of a disease which is almost peculiar to man, and results in muscles being in part turned into bone.

[1] *Trans. Roy. Dublin Soc.*, February 1896

The various views held by anatomists with regard to this most interesting find may be summed up as follows :

(1) That the remains—supposing them all to belong to the same individual—are those of an ape.

(2) That they are human, but represent a lower stage of humanity than anything previously discovered.

(3) That the bones represent a 'missing link.'

(4) That they are purely pathological, and represent a diseased human being.[1]

It is to be hoped that other and more complete remains will be discovered in Java, so that this highly interesting question may be solved in a satisfactory manner. The present doubt as to whether the thigh-bone belonged to the possessor of the skull is rather tantalising, the one being so much more human than the other. Then, again, the question of the geological age is of the greatest importance. Supposing the remains turn out to be Pleistocene! that alone would be almost fatal to the belief that the creature was not human, because in that, the latest geological period, man, as we know him now, *homo sapiens*, had fairly established himself.

We have entered somewhat fully into this case because it has attracted so much attention, and the reader will naturally wish to know the latest news with regard to the antiquity of man—*i.e.* how far back into geological time he can be traced.

At present, as far as we can see, it appears almost unquestionable that the Pleistocene represents the Ultima Thule, beyond which all is blank. But there is good ground for hope that this barrier to knowledge will at last be broken down, as so many other barriers have been, and that the 'ever-victorious army' of scientific workers will some day advance one more stage on their long

---

[1] *Nature*, 1895, and January 16 and 30, 1896; also *Natural Science*, 1894-6, and paper by Prof. Marsh (with photographs), *Amer. Jour. Science*, June 1896. The 'restoration' of the skull by Dubois, shown on p. 481 of this paper, is very ape-like.

# THE ANTIQUITY OF MAN

and arduous journey in search of truth. Every step gained is a step in advance, even if it is only a step in the apparently dry and barren region of hard fact, and leads that army on towards the source of all truth—

> On to the bound of the waste,
> On to the city of God.

---

NOTE.—A plaster cast of the calvarium from Java, and some other very old skulls, may be seen in the Natural History Museum, Cromwell Road (Geological Gallery, No. I., Case I.).

# Part II

# MEN OF THE LATER STONE AGE AND BRONZE AGE

## CHAPTER VIII

### DWELLERS ON THE WATER

*The antiquities of the first age (except those we find in Holy Writ) were buried in oblivion and silence: silence was succeeded by poetic fables: and fables, again, were followed by the records we now enjoy. So that the mysteries and secrets of antiquity were distinguished and separated from the records and evidences of succeeding times by the veil of fiction, which interposed itself, and came between those things which perished and those which are extinct.*—SIR FRANCIS BACON, Preface to *Wisdom of the Ancients*.

So wrote Francis Bacon in the seventeenth century! But the discoveries wrought since his days have made such a revolution that one feels tempted to exclaim with the Frenchman, 'Nous avons changé tout cela.' In most parts of the civilised world the relics of former ages have been brought to light, so that, to some extent at least, we can read their history. The broad outlines are beginning to appear, as when the photographer pours the fluid on to his sensitised plate. The 'development' of the details is only a matter of time. We propose in the present chapter to treat of that part of the story of primitive man which deals with the old lake-dwellers, whose homes were on the water, and constructed on piles, or otherwise. We shall find here a fairly complete picture. Let us first take the testimony of some old writers who actually saw the settlements of this kind.

Herodotus describes the lake-dwellings of the Pæonians as follows. That was 500 years or so B.C. 'Their dwellings are contrived after this manner: planks fitted on lofty piles are placed in the middle of the lake, with a narrow entrance from the mainland by a single bridge. These piles, that support the planks, all the citizens anciently placed there at the public charge; but afterwards they established a law to the following effect : whenever a man marries, for each wife he sinks three piles, bringing wood from a mountain called Orbelus; but every man has several wives. They live in the following manner : every man has a hut on the planks, in which he dwells, with a trap-door closely fitted in the planks and leading down to the lake. They tie the young children with a cord round the foot, fearing lest they should fall into the lake beneath. To their horses and beasts of burden they give fish or fodder, of which there is such an abundance that when a man has opened his trap-door he lets down an empty basket by a cord into the lake, and, after waiting a short time, draws it up full of fish.'[1]

Another old writer says : 'Concerning the people of Phasis, that region [which lies to the east of the Black Sea] is marshy and hot, full of water, and woody; and at every season frequent and violent rains fall there. The inhabitants live in the marshes and have houses of timber and of reeds constructed in the midst of the waters; and they seldom go out to the city or the market, but sail up and down in boats made out of a single tree-trunk, for there are numerous canals in that region. The water they drink is hot and stagnant, putrefied by the sun, and swollen by the rainfall, and the Phasis itself is the most stagnant and quiet-flowing of all rivers.'[2]

According to Pliny, the Chausi (Frisians and other races along the coast of the German Ocean) lived in houses built on artificial mounds surrounded at high tide by the sea. He says, in his well-known 'Natural History :' 'I have myself personally witnessed the condition of the Chauci, both the Greater and the Lesser, situate in the regions of the far North. In these climates a vast tract of land, invaded twice each day and night by the overflowing waves of the ocean, opens a question that is eternally proposed to us by Nature, whether those regions are to be looked

---

[1] Herodotus, *Terpsichore*, v. 14.   [2] Hippocrates, *De Aeribus*, &c., xxxvii.

upon as belonging to the land, or whether as forming a portion of the sea.

'Here a wretched race is found, inhabiting either the more elevated spots of land, or else eminences artificially constructed, and of a height to which they know by experience that the highest tides will never reach. Here they pitch their cabins; and when the waves cover the surrounding country far and wide, like so many mariners on board ship are they; when again the tide recedes, their condition is that of so many shipwrecked men, and around their cottages they pursue the fishes as they make their escape with the receding tide. It is not their lot, like the adjoining nations, to keep any flocks for sustenance by their milk, nor even to maintain a warfare with wild beasts, every shrub even being banished afar. With the sedge and the rushes of the marsh they make cords, and with these they weave the nets employed in the capture of the fish; they fashion the mud, too, with their hands, and, drying it by the help of the winds more than of the sun, cook their food by its aid and so warm their entrails, frozen as they are by the northern blasts; their only drink, too, is rain water, which they collect in holes dug at the entrance of their abodes; and yet these nations, if this very day they were vanquished by the Roman people, would exclaim against being reduced to slavery! Be it so, then. Fortune is most kind to many just when she means to punish them.'[1]

The custom of living in wooden houses on piles over the waters of a lake, river, or inlet of the sea is still practised by barbarous tribes, and has been described by many travellers in the Malay Archipelago, New Guinea, Venezuela, and in Central Africa. When Ojeda Vespucci and other discoverers entered the Lake of Maracaybo in 1499, they found an Indian village constructed on piles above the water, and therefore called it Venezuela, or 'Little Venice.' The Papuans of New Guinea build their dwellings of bamboo along the coasts and river banks. Captain

[1] Pliny, *Nat. Hist.*, Liber xvi. 1.

Cameron saw regular villages of pile-dwellings on Lake Mohrya, in Central Africa, each separate and accessible only by jealously guarded canoes. Lake-dwellings are also found in Lago Maggiore and on the Lake of Varese, and Garda in Lombardy; in Lake Salpi in the Capitanata, and in other parts of Italy. They have also been found in Austria and Hungary. Dwellings of this kind have long been known to the Japanese; in Dr. Keller's book will be found a drawing of lake-dwellings in Japan taken from a Japanese tea-tray, which was sent to the author by Dr. Henry Woodward. Turning again to Europe, we find sites in France. The little town of Berry is supposed to have been built on the site of a settlement. It is situated in the midst of a marsh since dried up. In the Jura Mountains they have been heard of; in the Pyrenean valley of Haute-Garonne, Ariège and Aude, as well as those of the Eastern Pyrenees, in the department of Landes, and on the lofty plain of Bearn are many marshy depressions where piles have been found.

In Great Britain pile-dwellings take the form of 'crannoges,' corresponding to the 'fascine structures' presently to be described. In England, according to Munro, they are known at the following places: Wretham Mere, near Thetford (Norfolk); Barton Mere, near Bury St. Edmunds (Suffolk); at Crowland, in the Fenland; in London; near Glastonbury (now being excavated); at Llangorse Lake, near Brecon, South Wales; in Berkshire, not far from Hermitage; Holderness (Yorks), and there are indications of probable lake-dwellings at other places in the county. Some of the British settlements will be described later on; at present we are merely concerned with the geographical distribution of lake-dwellings.[1]

[1] *The Lake Dwellings of Switzerland and Other Parts of Europe*, by Dr. F. Keller (2nd ed. 1878), is still the standard work on the subject. The reader should also consult the works of Sir J. Lubbock, Prof. Boyd Dawkins, and Prof. J. Geikie, already quoted in other chapters. Dr. Robert Munro's *Lake Dwellings of Europe* also is a very exhaustive work and beautifully illustrated. The following, when complete, will probably be the best work on the sub-

## DWELLERS ON THE WATER

Dr. Ferdinand Keller devoted all the later years of his life to the investigation of Swiss lake-dwellings. The reader will see from the following list where settlements have been found in Switzerland and Italy. On the great lakes they are generally opposite existing villages.[1]

LIST OF LAKE-DWELLING SETTLEMENTS[2]

| Lake of | Settlements |
|---|---|
| Constance | 24 |
| Nussbaumen | 1 |
| Pfäffikon | 5 |
| Greifensee | 2 |
| Zürich | 6 |
| Zug | 6 |
| Baldegg | 5 |
| Sempach | 6 |
| Wauwyl | 5 |
| Inkwyl | 1 |
| Burgäschi (canton Solothurn) | 3 |
| Mooseedorf | 2 |
| Bienne | 22 |
| Neuchâtel | 50 |
| Morat | 16 |
| Geneva | 24 |
| Luissel (near Bex) | 1 |
| Annecy | 2 |
| Bourget | 8 |
| Mergurago | 1 |
| Borgo Ticino | 1 |
| San Martino (piles not found) | 1 |
| Varese | 7 |
| Garda | 6 |
| Fimon (near Vicenza) | 2 |
| Castione (terra-mara beds) | 1 |
| Total | 208 |

---

ject—we have only seen one part so far—*Antiquités Lacustres* (Lausanne : Bridel et Rouge, 1894), the work of two Swiss societies assisted by the Government of Vaud.

[1] The subject is well illustrated at the museums of Neuchâtel, Berne, Constance, Zürich. Photographs, taken in 1875, of the piles at Lake Bienne may be seen in the Berne Museum. In that year the lake was lowered several feet.

[2] This list is drawn up from the accounts in Dr. Keller's book.

Lake-dwellings date back from very early times, for they were abundant in the Stone Age. In Switzerland they ceased soon after the introduction of iron, but in Britain the 'crannoges,' which are their equivalents, continued to be used as places of refuge far into historic times. The following list shows some of the principal settlements of each period. According to Munro, the settlements of the Stone Age are only found in a limited area in Central Europe, and chiefly in the lakes bordering both sides of the Alps. This area includes the lakes of Lombardy, Laibach, Bavaria, Switzerland, and Savoy, with the exception perhaps of Lake Bourget—where the settlements appear to belong to the Bronze Age only. But perhaps further discoveries may alter this conclusion.

*Stone Age*

Meilen, in the Lake of Zürich
Wangen . . . Constance
Robenhausen . . Pfäffikon (partially dried)
Mooseedorf . . Mooseedorf

*Bronze Age*

Morigen, in the Lake Bienne
Estaver . . . Neuchâtel
Morges . . . Geneva

*Iron Age*

La Tène (near Marin) Neuchâtel

The settlements of the Stone Age are, of course, distinguished by the entire absence of the metals, and by an abundance of implements, utensils, &c., of wood, bone, horn, flint and other stones. They are also nearer to the shore, the greatest distance being about 300 feet, whereas in the later Bronze Age they were about 1,000 feet or more distant in some cases. The piles of the Stone Age are thicker. The mere absence of metal would not be a safe test; but with the help of the others just mentioned it is not

Plate VI.

SWISS LAKE DWELLERS: AN EVENING SCENE
BRONZE AGE

## DWELLERS ON THE WATER

difficult to assign to a settlement its period in the prehistoric age. The pottery also is of a ruder and coarser make, with more sand in it, and with simpler types of ornament. There is no longer any question that *safety* was the chief object of the builders of these settlements. Here they could live secure from attack. Their sheep and cattle could not be stolen by night because they were driven every evening to their sheds near the huts. At the same time one can readily believe that in summer, at least, living over the water would be pleasanter than living on land.[1]

The history of research in this new field deserves to be briefly mentioned. As early as the year 1829 an excavation was made on the shore of Lake Zürich, opposite Ober Meilen, for the purpose of deepening the harbour, when piles and antiquities were discovered. Unfortunately no record of this kind was kept. But in the years 1853 and 1854, in consequence of an extraordinary drought and long-continued cold during the winter months, the rivers of Switzerland shrank so much that the level of the lakes was lower than it had ever been known to be before. The low shelving shores became dry land, and islands, never noticed before, appeared. The lowest level marked on the so-called Stäfa Stone was that of 1674, but in 1854 the water sank a foot lower. In that year Mr. Aeppli, of Meilen, on the lake, informed the society at Zürich of the discovery of the remains of human industry near his house, pointing out that they would be likely to throw some light on the early history of the country. Dr. Keller consequently began his now famous researches on the subject, and was the first to prove that the primitive people of his country lived on pile-dwellings.

Great numbers of piles, deers' horns, and implements were found by the inhabitants of Ober Meilen in a small bay between that place and Dollikon, who were adding to their

[1] The famous Blackmore Museum of Salisbury contains a beautiful little model of lake-dwellings, published by Max Götzinger in Basel. The British Museum contains a fine collection of lake-dwelling antiquities.

gardens by building a wall along the new water-line and filling in the extra space with mud from the shore in front. It was here that the first lake-dwelling came to light. The numerous objects of interest, now more or less familiar to all antiquarians, which have been discovered among the lake-dwellings, are usually found in the mud or peat, and often at considerable depths. But in some cases they are still lying uncovered on the bottom of the lake, after having been exposed for thousands of years to the gaze of every boatman who passed over them ! They can then be fished up with a large pair of forceps worked by a cord from a boat.

The walls consisted of wattlework woven in between the upright posts and covered over with clay. This wattlework is shown in our illustration (Plate VI), where a portion of the coating of mud is represented as having fallen off, so that we see the wattlework that was behind it. The roofs were thatched with straw or reeds from the lake. Every hut had its hearth, consisting of three or four large slabs of stone. Sometimes the stakes used for piles were hardened at their lower end and sharpened by the action of fire; at other times there is still clear evidence of their having been cut with axes of stone or of bronze. The marks made by the bronze chisels can even be distinguished from those made by stone axes. How the piles were driven in is not known. But in some cases the people found it easier to raise the bed of the lake round the piles with loose stones, &c.

But lake-dwellings were not all constructed on the same plan. We have the rude 'steinbergs' of the Stone Age, which were mere artificial heaps of stones and rubbish, the 'crannoges' of Ireland and Scotland, and the 'fascine structures' of Switzerland described by Keller. Instead of a platform supported on a series of piles, these latter erections consisted of layers of sticks or small stems of trees built up from the bottom of the lake till the structure reached above the water-mark, and on this series of layers the main platform for the huts was placed. In these dwellings upright posts were used as stays or guides for the great mass of sticks, reaching down to the bottom of the lake. Fascine structures occur chiefly on the smaller lakes and morasses, and appear to belong mostly to the Stone Age.

The construction of any kind of lake-dwelling, whether a true pile-dwelling, a fascine structure, or a crannoge, must have been a work of great labour; probably the services of a whole clan would be required. Crannoges appear to have been made somewhat in the following manner. First, a suitable place was chosen; this was generally a small mossy lake, its margin overgrown with grasses and weeds. All around was thick forest. Stones would only be used very sparingly on a site such as this, where the bed of the lake would be all soft oozy mud and decomposed vegetable matter. An island was made with stems of trees and brushwood, and these were pressed down with earth and stones.

A surrounding series of piles formed a stockade, and piles were also used to pin down the loose mass of material and keep it in place. In some cases the upright piles were held together by mortised beams, but generally by mere interlacing branches. Sometimes there were three concentric circles of piles round the artificially formed island; rows of piles were also fixed in lines radiating from the centre, so that altogether they formed a pattern like a spider's web. Some settlements were approached by a wooden gangway, others were provided with a stone causeway, which perhaps was slightly submerged, so that it could only be used by those who knew its direction. This is suggested by the fact that some of them had a zigzag direction and could only be waded by those who were intimately acquainted with its windings. In other cases the island could only be approached by boats.

Where a great depth of soft mud made it impossible for the piles to be firmly fixed, they were braced together by an arrangement of interlacing beams which served as tie-rods.

This plan was adopted at Wollishofen and other stations adjacent to the town of Zürich. Occasionally the piles are not round stems, but boards. The lowest bed within the enclosure is generally a mass of ferns, branches, and other vegetable matter, often covered with a layer of round logs cut into lengths of from four to six feet, over which comes a quantity of clay, gravel, and stones.

With regard to the general shape and external appearance of the houses themselves, it is clearly proved that, with one exception, they were square or rectangular, and therefore some of the 'restorations' of Troyons and others showing round houses are incorrect. Only one case is known of a circular dwelling. This is curious, because we know from the evidence of Trajan's column in Rome that the Germani used circular huts for dwelling-places on land, and the Roman soldiers are represented in the act of setting fire to them. The discovery of an actual lake-dwelling buried up in peat has added very much to our knowledge of the dwellings themselves. This discovery is thus reported by Dr. Robert Munro. 'For a long time the only indications that huts were erected over the platforms consisted of portions of clay having the impressions of round timber, hearthstones, and some stray beams and bits of thatching. Recently, however, more definite information has been brought forward by Mr. Frank, the investigator of the lake-dwelling of Schussenried. This settlement had none of the signs of having been destroyed by fire, and it is supposed that its inhabitants voluntarily abandoned it on account of the growth of the surrounding peat. In that case it is probable that the huts would be allowed to fall into natural decay; but before this happened there was a chance that some part of the building would become overtaken by the moss, and so become, as it were, hermetically sealed up. That something like this actually occurred is now proved by the discovery at the above station of the foundations and portions of the walls of a cottage deeply buried in the moss. Upon the discovery being known, Mr.

Frank had the ruins at once uncovered, and before the crumbling materials disappeared there was a plan of the building taken, which by the courtesy of the investigator I had an opportunity of inspecting. The structure was of an oblong rectangular form, about thirty-three feet long and twenty-three feet wide, and was divided by a partition into two chambers. The other, or inner chamber, was somewhat larger, and had no communication with the outside, except through the former by means of a door in the partition. There were no relics found in these chambers, but in the outer there was a mass of stones which showed signs of having been a fireplace. The walls were constructed of split stems set upright and their crevices plastered over with clay. The flooring in both chambers was composed of four layers of closely laid timbers separated by as many layers of clay. These repeated floorings may have been necessitated by the gradual rise of the surrounding peat, which ultimately drove the inhabitants away.'[1]

Since flint is not plentiful in Switzerland, we find the larger implements, such as axes, generally made of diorite, serpentine, and other hard stones, and even of jade. The presence of the latter stone is a matter of great interest, inasmuch as it probably was imported from the Far East. It therefore seems to bear witness to the fact that the lake-dwellers had commercial relations with other countries. Jade is not found in Europe, but occurs in China, India, and Egypt. This subject, however, is still rather a matter of controversy, for, though in spite of many inquiries, no site for native jade has been yet discovered in Europe, some authorities believe that the people found it somewhere in their own neighbourhood. It is certain, from the presence of chips in many places, that they worked it up themselves on the spot, and that gives some countenance to the idea. There are as many as 4,000 specimens of jade from Lake Constance alone. Two other minerals, known as jadeite and chloromelanite, closely resemble jade, and these are also found in the settlements,

[1] *Lake Dwellings of Europe*, by Dr. R. Munro, p. 508.

as well as in dolmens in Europe. A specimen of either jade (or jadeite) has lately been reported from a British barrow or tumulus. The boring of stone hammers was done by means of soft wood and sand, as described by Sir John Evans in his work on stone implements.[1] The axes were generally mere wedges fixed in a socket at the end of a short piece of stag's horn, which, again, was fastened into a handle of wood or horn.

Our knowledge of the state of affairs in the Swiss settlements during the Stone Age is chiefly derived from the excavations at Robenhausen. There is now a great deep turf moor where formerly were the waters of a lake. It is estimated that the platform on which Robenhausen was built was supported by 100,000 piles! These were originally about twelve feet long and eighteen inches in circumference. They were mostly of cedar, oak, and beech wood, and ran in regular lines. Cross-timbers were mortised on to the tops of these, and on the former a platform of beams was laid down to serve as a base for the houses. The dwellings, which we have already described, were usually one-storied huts and were closely crowded together. At Niederweil, near Robenhausen, two huts are said to have been unearthed by Mr. Messikommer. The same explorer found at Robenhausen what he considered to be indications of four separate dwellings, over an area of 99 by 30 feet. He concluded from the way the antiquities were grouped that each hut had a hearth, weaving machine, millstone, sharpening stones, &c. The area of each he reckoned at 750 square feet, which is almost identical with the area of those at Schussenried. Similar results were obtained at Irgenhausen. But at Niederweil the huts were smaller. Huts were also provided for the domestic animals belonging to these people, and doubtless, in the evening, sheep, goats, and cows were driven from the pastures on the shore into safe quarters in the town. Such a scene we have suggested in our illustration (Plate VI.), where cows are seen crossing the gangway and a man

[1] *Stone Implements of Great Britain*, p. 48.

is blowing a horn. Two other men are represented carrying a deer which they have killed in the chase.

Fishing, as might be expected, was much in favour, and specimens of fish-hooks, made of shell, bone, or bronze, are to be seen in the museums. At Robenhausen some long bows made of yew were found, one of which was over five feet long and very well preserved. Arrow heads, as well as spear heads, were made of flint, bone, or horn in the Stone Age settlements, and the former were fastened on with bitumen. Neat little saws fixed in wooden handles were made of flints with the edge carefully notched. Bone awls are frequently found, and these may have been used in making skins into garments. Boats were made out of the hollowed trunks of trees. Good examples of these occur in not a few places.

With regard to the clothing of the people we have hardly any evidence; but there can be little doubt that sheep skins were used for this purpose, especially in earlier times. It seems, however, equally certain that the people were able to weave both cloth and linen. As evidence that they were in the habit of spinning and weaving, there are spindle-whorls, clay weights and pirns, not to mention bits of thread and cord and fragments of coarse linen, both at Robenhausen and Wangen. Wild flax was used for the purpose. Dr. Keller says he has seen specimens of garments made of flax, but does not say if the specimens were complete. Like the ancient Britons, they were acquainted with wickerwork. Dressed leather has been found in some excavations, as well as a wooden last, showing that the people covered their feet with some sort of a leather shoe or sandal. Even in the Stone Age it is clear that the lake-dwellers were not insensible to the charms of personal ornament. Shells (both recent and fossil), coloured pebbles, the teeth of carnivorous animals, ornamented pieces of bone and horn, stone and clay beads, and even roundlets of the human skull were pierced for suspension, and worn either as pendants or as necklaces. M. Réclus, in his well-known work,

'The Earth,' mentions that in the year 1860 a slow-moving glacier in the Austrian Alps, which flows into the Ahrenthal, threw out a well-preserved corpse [1] still clad in a dress the ancient fashion of which had been abandoned by the mountaineers for centuries. In Irish peat-beds [2] bodies clothed in skins have also been discovered. Finds such as these may help the antiquarian better to picture the appearance of those who lived in lake-dwellings and crannoges. But the precise period they represent is unknown.

With regard to the boring of the stone hammers, it has been clearly shown that in some cases this was done with a tube of wood; for in some places hammer-stones have been found which were unfinished, and show a core of stone still left in the centre. Anyone can easily convince himself of the possibility of boring even hard stone with a piece of wood and some sand by trying the experiment. Blocks of stone in which grooves have been sawn have also been found, and in that case probably a thin strip of wood was used. At Robenhausen wooden dishes have been found which were cut out of the solid, as well as ladles and bowls.

Vessels of pottery, in very large numbers and of all sorts of shapes, have been found in settlements both of the Stone and Bronze Ages. And yet the potter's wheel was unknown! Everything was made by hand, though many are very perfect in form. The baking was very imperfect, having apparently been done on an open fire. Pottery of the Stone Age is very coarse and generally contains many grains of quartz. The ornamentation consisted of simple lines or of impressions made with the nail. Curved lines rarely occur, and no attempt was made to represent animals. One vase was found at Wangen on which was an attempt to represent a plant. Tumblers or drinking vessels

---

[1] Of the sixteenth century, according to Sir H. Howorth (*Glacial Nightmare*).

[2] See Sir W. Wilde's *Cat. Roy. Irish Ac. Museum*, p. 276.

were often made of pottery, and in the Bronze Age settlements these appear to have been supported on earthenware rings. Tables apparently were unknown in the earlier period, so that the vessels probably were placed on the floor.

A number of curious crescent-shaped things have been found, which may be of stone, wood, or clay, and flat at the base. Being in shape like half-moons, they have, until recently, been considered to be religious emblems because the moon was an object of worship. But they are now considered to be simply head-rests for the women—possibly for men as well. Hairpins are very plentiful in the old settlements, and may be of bone or bronze. Those of bronze are very artistic productions. It is clear that a woman could not very well sleep with a number of such pins in her hair unless her head was supported from off the ground, Probably the hair-dressing was of an elaborate character. Head-rests were used by ancient Egyptians, as visitors to the British Museum can see for themselves; and such are still used by the natives of Africa. Now, if the lake-dwellers worshipped the moon, there is no reason why they should not have made their head-rests into symbols of the crescent moon. And perhaps, in those days of superstition, the practice may have been adopted as a protection against evil spirits, or as a way of propitiating the moon goddess. The writer is inclined to think that these symbols, if such they are, have more resemblance to the horns of a cow; and as the cow was sacred in Egypt, Hathor being the cow-headed goddess, it is possible the lake-dwellers may also have used the cow's horns as a sacred symbol, probably in connection with the moon.

Fragments of small hollow globes, made in pottery, were found in a settlement of the Bronze Age at Möringen, which are believed to be fragments of children's rattles. Two of these, which are quite perfect, are now in the museum at Berne. They are ornamented, and contain inside a piece of hardened clay; when shaken they make a jingling noise. We may well believe that the important problem of how to amuse the baby and keep it from crying was one to which, by necessity, the lake-dwellers had to turn their attention. Let us hope that these rattles had the desired effect! Since there were no cupboards in early days, it is most likely that food was kept in earthenware jars. Ornaments appear to have been kept in similar receptacles.

In the earlier days armlets and necklaces were of bone or shell, but later on of bronze. Many of the bronze armlets, earrings, torque (or necklaces), and bracelets, brooches, &c., were most artistically designed and executed, and it must be remembered that all this bronze when worn was of a bright golden colour. Some of the hairpins were set with bright red coral or enamel or stones, and the amber necklaces must have been very pretty. The archæologists have much evidence with regard to the food of these people. From the first they possessed a certain elementary knowledge of agriculture. But as time went on, they became more skilled in that art, and devoted less time to hunting. In the Bronze Age settlements there are fewer remains of wild animals, and the domestic animals had been greatly improved by careful breeding. This is proved by the researches of Professor Rütimeyer. The people of the Early Stone Age had only one small species of dog, a small ox, a horned sheep, and the goat; while in the Bronze Age these were all larger, and in addition they had the horse. The latter was small and slender, judging from the bones found in not a few places, such as Möringen. Bridle-bits and horse-trappings are also found. The ass has also been recognised. The original *Bos primigenius* seems to have been tamed and crossed with an early type, giving rise to a variety of breeds. The dogs of the Bronze Age were larger. Domestic fowls and cats were then kept.

The following table [1] shows the proportions of wild and tame animals at Wauwyl and Mooseedorf, as representing the Stone Age, and Nidau the Bronze Age :—

[1] Taken from Lubbock's *Prehistoric Times*.

| Wild Animals | Wauwyl | Mooseedorf | Nidau |
|---|---|---|---|
| Brown bear | Several | Several | — |
| Badger | ,, | ,, | — |
| Marten | Common | ,, | — |
| Pine marten | ,, | ,, | — |
| Polecat | Several | ,, | — |
| Wolf | 1 | — | — |
| Fox | Common | Common | — |
| Wild cat | Several | Several | — |
| Beaver | ,, | Common | — |
| Elk | 1 | 1 | 1 |
| Urus | — | 1 | — |
| Bison | 1 | 1 | — |
| Stag | Very common | Very common | Very common |
| Roe deer | Several | Common | — |
| Wild boar [1] | ,, | ,, | — |
| **Domestic Animals** | | | |
| Domestic boar | 1 (?) | — | Common |
| Horse | Several | — | ,, |
| Ox | Very common | Very common | Very common |
| Goat | Several | Several | Common |
| Sheep | 1 | ,, | ,, |
| Dog | Several | ,, | ,, |

At Wangen, on the Lake of Constance, a great quantity of charred corn was dug out when the lake was low. Mr. Löhle believes he has obtained, at various times, as much as 100 bushels. Cakes of bread have been found made of grain roughly crushed on stone querns. In the ears from Wangen each row has generally ten or eleven grains, smaller and shorter than those now grown. Barley, millet, and peas were cultivated. Oats and the dwarf field-bean belong to the Bronze Age. Millet was also used for bread. The cakes made of wheat were flat and round, about one inch thick and four or five inches in diameter. Poppy-seed cakes were found at Robenhausen. Winter wheat, rye, and hemp

[1] The marsh boar, which is common, is considered by Prof. Rütimeyer to have been wild at first, but domesticated at Nidau and in the later settlements. Notice the absence of cats. One is inclined to wonder how the old maids got on without these pets! But perhaps there were no old maids—most likely not. The common fowl is also absent. Rats and mice have apparently left no remains.

are not found; nor are most of the ordinary vegetables now in use. Fruits and berries were used, but do not appear to have been cultivated. The following have been identified: apples, pears, plums, sloes, cherries, raspberries, blackberries, strawberries, hazel and beech nuts, chestnuts, poppies. It was at first doubtful whether the vine was cultivated, but grape-stones have been clearly identified by Professor Heer, both from Wangen and Steckborn.

It is interesting to note that the most ancient writings known to us agree with the testimony of the lake-dwellings. Thus flax is mentioned in the Pentateuch and in Homer, and was largely used by the Egyptians; but hemp, which has not been found in any of the settlements, is not mentioned. Neither oats nor rye are mentioned in the Book of Exodus, nor in Homer; and these are also absent.

M. Troyon, a former writer on the lake-dwellings, believed that the Bronze Age in Switzerland was inaugurated by a new race, who came and conquered the old Neolithic people; but this conclusion is not generally accepted. Dr. Keller thinks that the change took place gradually and peaceably, and that the same race continued on through the Bronze Age and even into the beginning of the Iron Age. It is interesting to note that the archæology of the lake-dwellings bears a striking resemblance to that of the graves, mounds (or barrows), cromlechs, or dolmens, of Scandinavia, France, and Britain; and this seems clearly to show that those who dwelt on the water were not a separate race. For example, the later pottery, ornaments, and implements of the lake-dwellers of the Bronze Age agree strikingly with those of the dolmens and round barrows. And thus the one department of archæology throws light upon the other.

It is possible, however, as Professor Virchow thinks, that during the Bronze Age the original lake-dwellers were joined by and mixed with a new race without any violent transition. A great deal has been made by Troyon and other writers of the frequent signs of the action of fire, and it has been hastily concluded that they testify to conquest and devastation by another race. But Keller believed that the traces of fire had been much exaggerated. Moreover, among villages where the houses were constructed of such inflammable material and with thatched roofs, it was only to be expected that they would frequently catch fire and so be destroyed, in spite of the abundant supply of water near. But then there were no means of 'turning it on.' At some time during the gradual transition from Bronze Age to Iron Age times there was a great conflict with the Roman Empire, and the old lake-dwellers of Switzerland were vanquished, never to appear again.[1] Superior knowledge and discipline, iron or steel weapons, and probably greater physical strength, were the main causes of this victory.

It is agreed by archæologists that in the Bronze Age pure and simple, when iron was almost unknown, the art of writing had not been discovered. Not a sign of writing exists on the French and other dolmens of the Bronze Age, nor in any relics from British round barrows. A new race, then, brought letters and iron.

'A battlefield at Tiefenau, near Berne, is remarkable for the great number of iron weapons and implements which have been found on it. Pieces of chariots, about a hundred swords, fragments of coat of mail, lance-heads, rings, fibulæ, ornaments, utensils, pieces of pottery and glass, accompanied by more than thirty Gaulish and Massaliote coins of a date anterior to our era, enable us to refer this battlefield to the Roman period. About forty Roman coins have also been found at the small island in the Lake of Bienne. After this period we find no more evidences of lake-dwellings on a large scale.'[2]

A number of objects of pure copper have been brought to light, and accordingly Munro and others consider that the Stone Age was followed by a transition period when pure copper was

[1] Outside this area the Iron Culture-Stage was well established.
[2] Lubbock, *Prehistoric Times*, 2nd ed. (1869), p. 213.

used and not bronze (which is an alloy of copper and tin). The copper implements are clearly imitations of the old flint weapons, which cannot be said of the bronze implements. This is especially the case with certain flat copper axes.

The lake village of Marin, in a little bay of the Lake of Neuchâtel, belongs to the Iron Age. As many as fifty wrought-iron swords were found here. They are handsomely ornamented and have bronze scabbards. The celebrated lacustrine station (so-called) of La Tène (the Shallows), which received this name from the fishermen on account of the former shallowness of the water, has been much discussed by antiquaries. It was formerly thought to be an ordinary lake-settlement, but of the Iron Age. A vast number of interesting and beautifully-wrought iron objects have been found there, together with numerous piles. On account of the lowering of the waters it was dry land in 1876, but no further explorations were made then. In 1880 M. E. Vouga, schoolmaster at Marin, began his investigations, and found a layer with Roman antiquities overlying another, which was Gallic. Wheels and other parts of waggons have been found here. The opinion that it was an ordinary settlement must be abandoned. Its geographical position, commanding the great highway between Constance and Geneva, and the fact that the greater number of its relics are warlike weapons, suggest that it was a military station. Many human bones were found here, and some skulls which had been cleft. Probably a fierce battle took place here, and the conquerors were Romans. But the majority of weapons, utensils, &c., are of a distinct type, and certainly *not* Roman. They mark a period called by Sir Wollaston Franks 'late Celtic,' and probably the people who made them were the Helvetii of Cæsar, who may have been a branch of the Celtic family. We might also call this period Early Iron Age.

Researches must be carried on a good deal further before it is possible for antiquarians to decide to what race, or races, the lake-dwellers belonged. The late Dr. Keller considered that they were all Celtic from the first, which, to say the least, is unlikely.

The barrows, chambered tumuli, and most ancient graves of Europe give clear evidence of an altogether different race which preceded the Celtic people, generally known as pre-Celtic or Iberian—probably of two preceding races. Hence it appears likely that the people who lived in those settlements which belong to the Stone Age may have belonged to the same early race (Neolithic) as those who made some of the older rude stone monuments, such as stone circles and chambered tumuli, a people whom antiquarians recognise by their long skulls and short stature.

Dr. Munro suggests that the original founders of the lake-dwellings in Central Europe were part of the first Neolithic immigrants who entered the country by the regions surrounding the Black Sea and the shore of the Mediterranean, and spread westwards along the Danube and its tributaries till they reached the great central lakes of Switzerland, where they established their settlements. Remains of lake-dwellings have been noted by travellers in Asia Minor.

Human bones and skeletons are by no means plentiful, and it was not until the year 1876 that any light was thrown on to the question as to how the people disposed of their dead. In that year some workmen discovered tombs at the pretty little village of Auvernier, on the Lake of Neuchâtel. The dead were not burned nor thrown into the lake, but buried in a stone cist. This tomb was built of slabs of granite, four of which were set upright to form the sides, and another laid flat on the top of them. In this coffin or cist fifteen to twenty skeletons were found. The bodies had been placed in a sitting posture, or, as antiquaries say, 'in the contracted position,' the knees being drawn up to the chin. They were placed round the tomb with their heads to the wall and their feet to the centre. Close by were two smaller stone chambers with bones. The relics found with the bones were a necklace of boars' tusks, and perforated teeth of the boar and wolf (probably for a necklace), a hatchet of serpentine, and two or three bronze rings, pins, and beads. A child had been buried close by without a coffin, but with bracelets and an amber bead. Amber generally marks the Bronze Age and Later Stone Age, as it was then an article of commerce, obtained by trading routes from the shores of the Baltic. Professor Rütimeyer pronounced these remains to be of the same type as those he had already examined from Nidau, Meilen, Robenhausen, and Wauwyl ; they are generally considered to belong to the *Early* Bronze Period—*i.e.* to a transition time from the Stone Age to the latter. They are of the *type de Sion*.

Other discoveries of human remains are reported from Montreux, some in stone cists, some without, and with a few bronze objects. Interments of the same age at Morges and St. Prex are described by Dr. Forel. At the latter place funeral urns with ashes were found, as well as interments. This is highly interesting as showing that both burial and cremation were practised at the same time. We shall see in the next chapter that the evidence of the British tumuli is in the same direction.

Burials of the true Stone Age were reported from Chamblandes, near Pully, and described by the late Morel-Fatio. Boars' teeth, some forty in number, all doubly perforated, lay about the chest of one skeleton : they evidently were worn for ornament, and may have been attached to the garments. Single skeletons were found always lying towards the east. Among other relics were perforated shells, portions of colouring matter (yellow and red), marine perforated shells, amulets made from human skulls, and doubtless worn as charms ; beads, apparently of coral, circular pebbles, a round hammer-stone, and a beautiful axe of serpentine. Other Stone Age burials occur not far off at Pierra-Portay and Châtelard.

We must now say a few words in conclusion on British crannoges. The following account describes the taking of an island fort in Ireland in the sixteenth century : ' There was one

Dualtagh O'Connor, a notorious traitor, that of all the rest continued longest as an outlaw, of power to do mischief. He had fortified himself very strongly, after their manner, in an island or crannoge within Lough Lane, standing within the County of Roscommon and on the borders of that country called Costelloghe. A few days ago, as opportunity and time served me, I drew a force of a sudden one night and laid siege to the island before day, and so continued seven days, restraining them from sending any forth or receiving any in, and in the meantime I had caused divers boats from Athlone and a couple of great iron pieces to be brought against the island, and on the seventh day we took the island, without hurt to any on our side, save my brother John, who got a bullet wound in the back. When our men entered the island there was found within it twenty-six persons, whereof seven were Dualtagh's sons and daughters; but himself and eighteen others, seeking to save themselves by swimming, and in their cot to recover the wood next the shore, were for the most part drowned. Some report that Dualtagh was drowned, but the truth is not known. It was scarce daylight, and the weather was foggy when they betook themselves to flight. The Irishry held that place as a thing invincible.'[1]

The late Dr. J. Robertson, speaking of Scottish crannoges, says: 'Among the more remarkable of the Scottish crannoges is that in the Loch of Forfar, which bears the name of St. Margaret, the Queen of King Malcolm Canmore, who died in 1097. It is chiefly natural, but has been strengthened by piles and stones, and the care taken to preserve this artificial barrier is attested by a record of the year 1508. Another crannoge—that of Lochindorb, in Moray—was visited by King Edward I. of England in 1303, about which time it was fortified by a castle of such mark that, in 1336, King Edward III. of England led an army to its relief through the mountain passes of Athole and

[1] Sir R. Bingham to Burghley, Dec. 16, 1590, in *Calendar of State Papers of Ireland*, vol. clvi., p. 374.

Badenoch. A third crannoge—that of Loch Cannor or Kinord, in Aberdeenshire—appears in history in 1335, had King James IV. for its guest in 1506, and continued to be a place of strength until 1648, when the estates of Parliament ordered its fortifications to be destroyed. It has an area of about an acre, and owes little or nothing to art beyond a rampart of stones and a row of piles. In the same lake there is another and much smaller crannoge, which is wholly artificial. Forty years after the dismantling of the crannoge of Loch Cannor, the crannoge of Loch an Eilan (Loch of the Island), beautifully situated among the mountains, in Strathspey, is spoken of as "useful to the country in times of troubles or wars, for the people put in their goods and children there, and it is easily defended."[1] Canoes hollowed out of the trunks of oaks have been found, as well beside the Scotch as beside the Irish crannoges. Bronze (brass) vessels, apparently for kitchen purposes, are of frequent occurrence, but do not seem to be of a very ancient type. Deer horns, boars' tusks, and the bones of domestic animals have been discovered; and in one instance a stone-hammer, and in another what seemed to be pieces for some such game as draughts or backgammon, have been dug up.'[2]

[1] 'An Account of Some Remains of Antiquity in Forfarshire' (Queen Margaret's Inch, Loch of Forfar), *Arch. Scot.*, vol. ii.
[2] The literature of lake-dwellings is already considerable. Dr. Munro, from whose valuable and exhaustive work we have taken much of the information here given, appends a list of 469 books and papers on the subject, mostly French, German, and Italian.

## CHAPTER IX

ABODES OF THE LIVING AND THE DEAD

In these shows a chronicle survives.—WORDSWORTH

IN the last chapter we followed the peoples of the Later Stone Age, or Neolithic period, and the succeeding Bronze Age to their homes on the water, of which such ample and well-preserved records are found on the borders of the Swiss lakes and elsewhere. It goes without saying that even in those early times lake-dwellings were not the only kind of human habitation in use, although they had some obvious advantages, such as security from attack, and food supplies in the form of fish, very handy, to say nothing of the cool, refreshing breezes in summer. But lakes are not to be found everywhere, and therefore it follows as a matter of certainty that the Neolithic and Bronze Ages folk, who were fairly numerous and widespread, must have had many settlements on the land.

The present chapter will deal with these people or peoples as they were on land, their daily lives, their habits, customs, religion, &c., in fact all that is known about them. And here we find the conditions exactly reversed: for whereas our knowledge of the lake-dwellers is derived from their actual homes, which tell us so much about their daily life, what we know of the land-dwellers of these times is chiefly derived from their tombs. The one picture is a scene of life, the other a scene of death. The pictures are, of course, far from complete, but the evidence derived from archæological discoveries is at least sufficient to enable us to give, as it were, outline sketches. The details can only be filled in gradually by future generations.

## THE MEN OF THE LATER STONE AGE

But, in the first place, it will be necessary to say a few words about a primitive race of people who are not known to us either by huts or tombs, or any kind of human habitation, but only by certain great heaps of shells and other *débris* which form the only records left to us. These are the 'kitchen middens' of Denmark, generally met with near the coast, and principally on the shores of the Lymfjord and the Kattegat. The mounds, or banks, are in some cases only three to five feet in height, in others they reach a height of ten feet. In width they are 150 to 200 feet, with a length sometimes of nearly 350 yards. In the case of a low and shelving shore they are only a few feet above high-water mark, but where the coast is steep they are higher.[1]

In some instances these shell-mounds occur as far as eight miles from the present coast, but in such places it is probable that a good deal of fresh land has been formed in the usual way, by accumulation of sediment brought down by rivers. Along the west coast of Denmark they are entirely absent, but the sea has made great encroachments there, so that any shell-mounds which may have existed have been swept away by the waves. The most abundant shells are those of the oyster, cockle, mussel, and periwinkle. The people evidently lived largely on shell-fish, hence we cannot expect to find traces of them inland.

In many places hearth-stones were discovered, arranged in such a manner as to form small platforms, and bearing all the marks of fire.[2] Those mounds which have been excavated disclose layers of charcoal, thus bearing out the evidence of the hearth-stones.

---

[1] Many of these deposits have been carefully examined, and thousands of specimens are deposited in the fine museum at Copenhagen. The results, obtained by a committee consisting of Prof. Steenstrup, Prof. Forchhammer, and Prof. Worsaae, are contained in six reports in Danish presented to the Academy of Sciences at Copenhagen. M. Morlot has published an excellent abstract of these reports in the *Mémoires de la Société Vaudoise*, tome vi., 1860.

[2] The name applied to the shell-mounds is derived from the Danish words kjökken, 'kitchen,' and mödding, a 'refuse-heap,' which word corresponds to our local term 'midding.'

But, according to Lubbock, it appears that in some places the people cooked their meals on the shore, so that shells and bones are mixed up with sand and gravel. They probably lived in tents or huts, and there is reason to think that these were erected where certain hollows or depressions are now visible on the mounds. At Havelse there is a small midden in the form of a ring enclosing a space where several dwellings may have been erected.

The cockle, mussel, and periwinkle shells are larger than those of the same molluscs that now live upon the coasts; and the oyster, which formerly was so abundant, has quite disappeared. Lyell concluded from these facts that the ocean had freer access than now to the Baltic, communicating probably through the peninsula of Jutland, which, at no very remote period, was an archipelago.

Weapons and implements have been found in great abundance, all made of bone, horn, shell, or stone, and it appears to be fairly established that the people who lived here were entirely ignorant of metal. The implements, &c., are roughly made, and hardly ever polished. Rude axes of the now recognised 'shell-mound type' have been found, together with flint-flakes, arrow-heads, sling-stones, and numerous fragments of bone.

The exact period to which these relics belong has not yet been settled. Some authorities, judging from their rude make, would place them quite at the beginning of the Neolithic period; others think that more civilised people were living inland, who were capable of turning out well-ground and polished weapons, such as occur in great numbers in the peat-beds of Denmark and Scandinavia, and so consider them Late Neolithic. One thing is clear, namely, that they cannot be Palæolithic, for in the Older Stone Age Europe was inhabited by a fauna now largely extinct, as shown in previous chapters. Also the physical geography of Europe was quite different, so that there were no lines of coast where the shell mounds occur.

The entire absence of metal gives to archæologists good

reason to believe that they belong to the Stone Age; and, although this important question of the age of the shell-mounds is not yet settled beyond dispute, yet there are other evidences that support the above conclusion. At all events, they represent the Neolithic state of culture, *even* if they lived at a time when bronze was coming in elsewhere. Certain relics, such as dug-out canoes found low down in the old peat-bogs of Denmark, are believed to be of the same age, viz. Neolithic. But we must always remember that these so-called Ages of Stone and of Bronze largely overlapped each other. There was no such thing as a universal Bronze Age.

The fauna of the shell-mounds forms in itself a valuable record of the people who lived on them, telling us their habits and daily lives. Thus we find they had only one domestic animal, and that was the dog, a fact which implies a great deal. They were essentially hunters, not agriculturists, and no trace occurs of any other domestic animal, such as the ox, sheep, goat, or pig. The bones of the stag and wild boar are very abundant, but the reindeer is probably absent altogether, although Steenstrup affirms that such is not the case. Remains of the wild bull (*Bos urus*) were so frequently found as to lead one to suppose that it was a favourite food with these people. This fierce animal was seen by Julius Cæsar, and survived long after his time. The aurochs, or bison, still preserved by the Tsars of Russia in the Lithuanian forest, has not yet been noted here.

The fish remains include those of the herring, dorse, cod, flounder, eel, and there are also bones of several birds. Shell-mounds of various ages occur in Scotland up to nearly fifty feet above sea-level; these may be seen on the shores of the Firth of Forth, of St. Andrew's Bay, of Forfarshire, of the Moray Firth, and in the Outer Hebrides, as well as in Ireland. They are also found in North America. There are some wretched people of a low type living on the coast of Tierra del Fuego, who feed principally on shell-fish, and are ignorant of agriculture. The

## ABODES OF THE LIVING AND THE DEAD 195

following account of them from Darwin [1] helps us the better to picture the state of the people who once lived in a similar manner on the shores of Denmark :—

'The inhabitants, living chiefly upon shell-fish, are obliged constantly to change their place of residence; but they return at intervals to the same spots, as is evident from the pile of old shells, which must often amount to some tons in weight. These heaps can be distinguished at a long distance by the bright green colour of certain plants which invariably grow on them. . . . The Fuegian wigwam resembles in size and dimensions a haycock. It merely consists of a few broken branches stuck in the ground, and very imperfectly thatched on one side with a few tufts of grass and rushes. The whole cannot be so much as the work of an hour, and it is only used for a few days. . . . At a subsequent period, the "Beagle" anchored for a couple of days under Wollaston Island, which is a short way to the northward. While going on shore, we pulled alongside a canoe with six Fuegians. These were the most abject and miserable creatures I anywhere beheld. On the east coast the natives, as we have seen, have guanaco cloaks, and on the west they possess sealskins. Amongst the central tribes the men generally possess an otter skin, or some small scrap about as large as a pocket-handkerchief, which is barely sufficient to cover their backs as low down as their loins. It is laced across the breast by strings, and, according as the wind blows, it is shifted from side to side. But these Fuegians in the canoe were quite naked, and even one full-grown woman was absolutely so. It was raining heavily, and the fresh water, together with the spray, trickled down her body. . . . These poor wretches were stunted in their growth, their hideous faces bedaubed with white paint, their skins filthy and greasy, their hair entangled, their voices discordant, their gestures violent and without dignity. Viewing such men, one can hardly make oneself believe they are fellow-creatures and in-

[1] *Journal*, p. 234.

habitants of the same world. . . . At night, five or six human beings, naked, and scarcely protected from the wind and rain of this tempestuous climate, sleep on the wet ground coiled up like animals. Whenever it is low water, they must rise to pick shell-fish from the rocks; and the women, summer and winter, either dive to collect sea-eggs, or sit patiently in their canoes, and, with a baited hair-line, jerk out small fish. If a seal is killed, or the floating carcase of a putrid whale discovered, it is a feast; such miserable food is assisted by a few tasteless berries and fungi. Nor are they exempt from famine, and, as a consequence, cannibalism accompanied by parricide.'

We pass on now to the tumuli, or mounds of earth, which mark the burial-places of the Indo-European and other races. In Greece there are plenty of them. They were called κολωνοί (hills) and also χώματα (heaps). Of this kind are the enormous mounds of earth still to be seen on the shores of the Hellespont, which, according to the Greek tradition, contain the remains of Homeric heroes like Achilles, Patroclus, and others. Tombs of this kind were erected by the Athenians in the plain of Marathon to those who fell there, the largest of which was originally thirty feet high. Frequently they were surrounded by a stone enclosure or wall. According to Diodorus, Semiramis, the widow of Ninus, buried her husband within the precincts of the palace, and raised over him a great mound of earth. Pausanias mentions that stones were collected together, and heaped up over the tomb of Laius, the father of Œdipus. Virgil says that Dercennus, king of Latium, was buried under an earthen mound; and, according to the earliest historians, whose statements are confirmed by the researches of archæologists, mound-burial was practised in ancient times by the Scythians, Greeks, Etruscans, Germans, and many other nations. The Chinese had a similar custom—one which is a legacy from the Stone Age—and some huge mounds of great antiquity are the resting-places of their ancient kings. Many of the great mounds of North America are undoubtedly sepulchral, while others cannot be so explained. In Egypt the mound evolved into a pyramid.

Homer's account of the building of the barrow of Hector brings the whole scene vividly before us. For nine days wood was collected and brought in carts, drawn by oxen, to the site of the funeral pyre. Then the pyre was built and the body laid upon it. After burning for twenty-four hours, the smouldering embers were extinguished with libations of wine; friends then picked the bones out of the ashes, and placed them in a metallic urn, which was deposited in a hollow grave, or cist, and covered over with large well-fitting stones. Lastly, a huge barrow, or tumulus, was heaped up over the grave, and then the funeral feast was celebrated.

The tumuli in the Troad, like the 'long barrows' of Wiltshire, were placed in commanding positions, with the idea of rendering them conspicuous. A reference to Homer's 'Iliad' (vii. 84–90) will prove this. Hector challenges one of the Greeks to meet him in single combat. If Hector should prove victorious, he undertakes to give back the body of his enemy to the Greeks, so that they may make for him a tumulus on the shores of the Hellespont, and that someone may say in times long afterwards, as he sails on the sea—'Yonder is the mound of some man who died long ago, and who was slain by Hector, when he was showing valour in the fight.' A stone pillar (the prototype of our column) was often placed in a mound, in order to make the spot conspicuous; such was the στήλη. It was often of large size. In the 'Iliad' (xi. 370) we read that Paris, who was skilled in archery, takes his position behind the pillar on the mound of Ilus, and shoots at Diomede, wounding him in the foot. This shows that it was large enough to conceal and protect a man. The Menhir, or standing stone, often marks the site of a burial in various parts of the world, but is sometimes only a memorial. We shall have more to say about these interesting landmarks in Chapter XI.

Homer ('Iliad,' xxiii. 166) also tells of the funeral of Achilles, and the details in this description show a striking agreement with the customs observed by those who constructed the European mounds in the days of the Bronze and the Iron Ages. The body, he says, was brought to the pile in an embroidered robe, and jars of unguents and honey were placed beside it. Sheep and oxen were slaughtered at the pile. The bones were collected from the ashes and placed in a golden urn along with those of Patroclus, the hero's dearest friend. Afterwards, a great mound was raised on the high headland, so that it might be seen from afar by future generations of men.

It is as well to make full use of all the historical notices of burials in old days, because they help to throw light on archæological discoveries, and explain details, the meaning of which might otherwise escape our notice. Herodotus, describing the funeral customs of the Scythians, states that, on the death of a chief, the body was placed upon a couch in a chamber sunk in the earth and covered with timber, in which were deposited all things needful for the comfort of the deceased in the other world. One of his wives was strangled and laid beside him, his cup-bearer and other attendants, his charioteer, and his horses, were killed and placed in the tomb, which was then filled up with earth, and an enormous mound raised high over all. The numerous barrows of the plains of ancient Scythia confirm the above description.

Julius Cæsar[1] says of the Gauls: 'Their funerals are magnificent and sumptuous. Everything supposed to have been dear to the deceased during his life was flung upon the funeral pile; even his animals were sacrificed, and, until quite recently, his slaves and the dependants he had loved were burned with him.' He also says that if the circumstances of the death of a chief were suspicious, the wives were put to severe torture and killed by fire ('Bell. Gall.' vi. 19).

The practice of burying in barrows was continued down to and even into the Christian era, but the greater number of tumuli are

[1] *De Bello Gallico*, iii. cap. ii. 2.

prehistoric. There are a few of which the date is known, such as those of Queen Thyra Danebod and King Gorm, who died about 950 A.D. at Jellinge, in Denmark. It is a mound about 200 feet in diameter and over 50 feet high, containing a chamber 23 feet long, 8 feet wide, and 5 feet high. This chamber, unlike those of the Stone or of the Bronze Age, is formed of massive slabs of oak instead of stone. In the Middle Ages it was plundered, but a few valuable relics were found when it was reopened a few years ago, among which was a silver cup. Old pagan customs were still kept up to a great extent after the introduction of Christianity ; and King Harald, the son and successor of old Gorm, who is said to have christianised all Denmark and Norway, erected a chambered tumulus over the remains of his father, on the summit of which was placed a rude pillar-stone, bearing on one side the memorial inscription in Runes, and on the other a representation of the Saviour of mankind ! The king's hows (barrows) at Upsala, in Sweden, known as the tumuli of Odin, Thor, and Freya, rival those of Jellinge in size and height. One of these, opened in 1829, was found to contain an urn full of calcined bones, together with gold ornaments recognised by their workmanship as belonging to the fifth and sixth centuries of our era.

Tumuli are sometimes mentioned in old records. Thus Gregory of Tours tells a story to the effect that Macliar, flying from his brother Chanaon, took refuge with Chonomor, Count of the Bretons. Chanaon sent messengers to demand that Macliar should be given up to him ; but Chonomor concealed him in a tomb, ' rearing over him a tumulus in the usual manner, but leaving a small opening for the entrance of air.' He then showed this tumulus to the messengers, and assured them that Macliar was buried in it !

The Danish sagas relate that in the middle of the eighth century Sigurd Ring, having conquered his uncle, King Harold Hildetand, in the battle of Braavalla, washed the corpse, placed it on Harold's war chariot, and buried it in a tumulus which he had formed for the purpose. Harold's horse was slain and buried with him, with the saddle, so that Harold might either ride to Valhalla, or go in his chariot, as he preferred. Ring then gave a great feast, after which he recommended the chiefs present to throw their ornaments and arms into the tumulus in honour of Harold. Finally, the tumulus was carefully closed.

In the later time of the Vikings, burial without cremation was practised. Some of the larger Viking barrows contained the ship fully equipped as she rode to sea, and the owner laid in state in a house constructed on the deck, as in the case of the Viking ship discovered in 1880, at Gokstad Sandefjord, and now at Christiania.

With many ancient peoples it was the custom to make the tombs of the dead like the abodes of the living. Among the early inhabitants of Italy—Etruscans and others—this is well illustrated, as proved by the singular cinerary urns found in the necropolis of Alba Longa, which are obvious imitations of rude huts formed of boughs and covered with skins. In the paintings on the walls of Etruscan tombs we see the style of the internal decorations of their houses.

The large and striking examples of early Lycian tombs in the British Museum show clearly enough that the whole structure was a copy or imitation of a wooden building, and the rafters have been copied in stone. Sometimes the walls were panelled, and weapons suspended therefrom, and easy arm-chairs with footstools attached were all carved out of the solid rock of the subterranean tombs. The hut-urns of the Swiss lake-dwellings also are imitations of rude huts made with boughs of trees.[1]

The origin of the art of architecture is perhaps to be found in the endeavours of early man to provide for his ordinary wants, such as rest, comfort, and shelter from the elements. Vitruvius gives a picturesque account of the early stages in the evolution of architec-

---

[1] See Lubbock, *Prehistoric Times* (1869), p. 51.

## ABODES OF THE LIVING AND THE DEAD 201

ture, and points out that man in his first state began to imitate the nests of birds and the lairs of beasts, and so constructed arbours with twigs and branches of trees. As an improvement on these he then invented huts with walls composed of dried turf, strengthened with reeds and branches. Other writers have endeavoured to trace three orders of primitive dwellings, the cave, the hut, and the tent, constructed severally by the tribes who devoted themselves to hunting and fishing, to agriculture, and to a nomadic life. Climate and other circumstances of course affected the form of the primitive building, as well as the material employed.

In Scandinavia, instead of chambered cairns, long chambered barrows, and dolmens, we find the well-known 'passage-graves,' or 'gang-gräber' (Ger.), which often contain bodies in the contracted position. They appear to have been used both as habitations for the living and tombs for the dead, and belong to the Later Stone Age. They contain a passage, formed of great blocks of stone, almost always opening towards the south or east, never to the north, and leading into a large central chamber, round which the dead are placed, often in a sitting posture. It is a striking and important fact that the dwellings used by Arctic nations—namely, the winter-houses of the Esquimaux and the 'yourts' of the Siberians—show a close resemblance to the 'passage-graves.'

The Siberian yourt consists of a central chamber sunk a little in the ground, and formed of stones (or sometimes of timber), while earth is heaped up on the roof and against the sides, thus making it outside like a mound. The opening is to the south. According to Captain Cook, the winter-houses of the Tschutski, in the extreme north-east of Asia, are very similar, being 'exactly like a vault.'

The inhabitants of the Aleutian Isles construct large half-underground habitations, often of great size. They are divided into compartments, one for each family. As people died, they were buried in these compartments, the body being doubled up into the 'contracted position.' The Laplander's hut, or gamme, also shows the same correspondence with a 'passage-grave.'

It was Professor Sven Nilsson, the venerable Swedish archæologist, who suggested that the passage-graves are a copy of the dwelling-house. In fact, it may even be said that in some cases they actually were the dwelling-house. It will be worth our while to follow him in his ingenious theory that the first houses of early man were imitations of a yet more primitive dwelling-place, namely, the cave. He reasons thus : [1]

Primeval man would be forced to seek for shelter. This he would find in mountain caverns, where the cold of night and the heat of day would be avoided. All the oldest traditions refer to this fact. The earliest inhabitants of Greece dwelt in mountain caverns. The people in Siberia who preceded the Samoyedes lived in subterranean abodes. Diodorus of Sicily (or D. Siculus) expressed the same idea in the words '*hieme in speluncas refugere.*' The Cyclopes of Homer,[2] although endowed by him with impossible attributes, are in reality a remnant of an older race belonging properly to the Stone Age, while the men of whom the great bard sang so sweetly were living in the Bronze Age.

The country between the Black and the Caspian Seas must be regarded—so the philologists say—as a region from which came a race capable of high civilisation, and European traditions also point to that region. The nations of the south and east in the Old World buried their dead in the same kind of habitations as those in which ages ago they themselves had dwelt.

Thus the Hittites buried in mountain caves, and Abraham (Genesis xxiii.) bought from them a double cave in which to bury Sarah, and long afterwards the Jews were accustomed to bury in caves and crypts. Professor Nilsson imagines that when the

---

[1] *The Primitive Inhabitants of Scandinavia . . . during the Stone Age.* Edited by Sir John Lubbock, F.R.S., 1868.
[2] *Odyssey*, Book I., 113–115.

cave-dwellers of the Caucasus were driven out by more powerful hordes and retreated to countries where there were no mountains, they took to making, as it were, artificial caves of stones, or even timber. In this way, he concludes, arose architecture.

In more northern regions the old savages sought out and partly made mountain caves with a passage or gallery pointing towards the sun, and even animals have the same instinct. There is abundant evidence that caves all over Europe were, in the Later Stone Age, inhabited by human beings. But if these old savages wandered from the mountains into plains in order to live there, they were obliged to collect blocks of stone, and to form with them artificial caves, or underground dwellings. The gallery-house was made thus, and became a reminiscence of the mountain cave, and we find the galleries, or passages, always pointing south.

Speaking of the old Turanian race, Canon Isaac Taylor says:—

'The vast and numerous monuments which constitute the tombs of this race can always be recognised; they exhibit a most remarkable and significant unity of design and purpose. These tombs are all developments of one hereditary type; they are all the expression of one great hereditary belief, and they all serve the purpose of one great hereditary cultus. The type on which they are modelled is the house. The belief which they express is the fundamental truth which has been the great contribution of the Turanian race to the religious thought of the world— the belief in the deathlessness of souls. The cultus which they serve is the worship of the spirits of ancestors, which is the Turanian religion. The creed of the Turanians was "Animism." They believe that everything animate or inanimate had its soul or spirit; that the spirits of the dead could still make use of the spirits of the weapons, ornaments, and utensils which they had used in life, and could be served by the spirits of their slaves, their horses, and their dogs, and needed for their support the spirits of those articles of food on which they had been used to feed. Hence, when we open these ancient Turanian sepulchres, we find that the resting-places for the dead have been constructed on the exact models of the abodes of the living; the dead have been carefully provided with the necessaries of life—the warrior is buried with his spears and his arrows, the woman with her utensils and her ornaments; by the side of the infant's skeleton we find the skeleton of the faithful house-dog, slaughtered in order that the soul of the brave and wise companion might safely guide the soul of the helpless little one on the long journey to the unknown land. In all respects the tomb is the counterpart of the house, with the sole difference that it is erected in a manner more durable and more costly. The Turanian tombs are family tombs; the dead of a whole generation are deposited in the same chamber.'[1]

> Here bring the last gifts! and with these
> The last lament be said;
> Let all that pleased, and yet may please,
> Be buried with the dead.
>
> Beneath his head the hatchet hide,
> That he so stoutly swung;
> And place the boar's fat haunch beside,
> The journey hence is long!
>
> And let the knife new sharpened be
> That on the battle day
> Shore with quick strokes—he took but three—
> The foeman's scalp away!
>
> The paints that warriors love to use
> Place here within his hand,
> That he may shine with ruddy hues
> Amidst the spirit-land.
>
> Schiller, *Nadowessische Todtenklage*, translated by Lord Lytton.

The tumuli of Great Britain and Ireland have been the subject of diligent inquiry for many years.[2]

---

[1] *Etruscan Researches*, by Canon Isaac Taylor, p. 36.

[2] The most recent and comprehensive work on the subject is *British Barrows*, by Canon Greenwell, of Durham, who has himself opened and reported on 232 barrows in Yorks. Aubrey, Sir Richard Colt Hoare, Stukeley, Dr. Thurnam, and others have also written accounts of their researches. Lieut.-

Camden long ago expressed the popular idea in saying that 'these burrows, or barrows, were probably thrown up in memory of soldiers slain thereabouts . . . because bones are found in them.' But Stukeley, later on, was much nearer the mark when he wrote : 'At Stonehenge one may count round about it forty-five barrowes. I am not of the opinion that all these were made for burying the dead that were slain herabout in battels ; it would require a great deale of time and leisure to collect so many thousand loades of earth ; and the soldiers have something els to doe flagrante bello : to pursue their victorie, or preserve themselves pursued : the cadavera remained a feast for the kites and foxes. So that I presume they were the mausolea or burying places for the great persons and rulers of those times. They are assuredly the single sepulchres of kings and great personages buried during a considerable space of time, and that in peace. There are many groups of them together, and as family burial places ; the variety of them seems to indicate some note of difference in the persons there interred, well-known in those ages.'

The classification of the Wiltshire barrows, and of the objects found in them, is due to the late Dr. Thurnam. His two valuable contributions to the Society of Antiquaries have been published by that society.[1] He divided them into long barrows, long-chambered barrows, oval barrows, and round barrows, which last may be disc-shaped, or bell-shaped, or bowl-shaped.

Long barrows are generally immense mounds, varying in size

General Pitt-Rivers (formerly Col. Lane Fox) has very systematically explored both camps and barrows in Wilts and Dorset, and published works on the subject copiously illustrated (*Excavations on Cranborne Chase*, &c., also papers in the *Archæologia* and *Jour. Anthropological Institute*). The beautiful models made to scale in his fine museum at Farnam (Dorset), on his own estate, are unique, and the writer greatly enjoyed inspecting them under the General's guidance. Visitors to Salisbury, Poole, and those parts are recommended to visit this most instructive collection, and the beautiful Larmer grounds, all of which this munificent archæologist has generously provided for and opened to the public.

[1] *Archæologia*, vols. xlii. and xliii.

from 100 or 200 feet to 300 and even nearly 400 feet in length; from 30 to 50 feet in breadth or upwards, and from 3 feet to 10 feet or 12 feet in height. They are surrounded by a deep wide trench, from which the material of the mound was taken, but there was always a break in the trench, probably for the entrance to the sepulchral chamber. These mounds are often placed east and west, the east end being rather higher and broader, and the sepulchral deposit is usually found under this more prominent end at or near the natural level of the ground. On the other hand, about one in six of the Wiltshire barrows is placed nearly north and south.

Long barrows mostly occur in the south-western counties, Wilts, Dorset, and Gloucestershire. But there are some in Caithness. They often contain a stone chamber of megalithic structure, with a passage leading to it from the outside. There are various forms of chamber; sometimes it is cruciform, at others it is divided into a series of niches, or takes the form of a long passage. Sometimes the chambers are quite apart from each other. For descriptions the reader is referred to Fergusson's 'Rude Stone Monuments.' But such structures are not seen in localities where there is no suitable supply of large blocks of stone. An enclosing wall of loose stones without mortar of any kind is sometimes seen, as at Uley barrow, in Gloucestershire (see Plate VII.). There is reason to believe that originally all, or most of, the long barrows were surrounded by a ring of large upright stones.

> They marked the boundary of the tomb with stones,
> Then filled the enclosure hastily with earth.
> *Iliad*, xxiii. 255.

In the case of the West Kennet barrow (as restored by Dr. Thurnam) there was a boundary wall of stone from two to three feet high, with large upright blocks of stone placed at intervals, forming a peristyle like those surrounding the 'topes' of India; and it has been pointed out that, according to Aristotle, the

Plate VII.

AN INTERMENT IN A LONG BARROW
LATER STONE AGE

Iberian people were in the habit of placing as many obelisks round the tomb of the dead warrior as he had slain enemies ; and it is not without interest that a structure of this sort has been noticed in Britain, because there is good reason to believe that the Neolithic people of Britain were the long-headed Iberians of small stature. But we shall refer to this subject again later on.

In Scotland the time-honoured cairn assumes the foremost place among all sepulchral memorials, and seems to take the place of the barrows, tumuli, or mounds met with in other parts of Britain. In many districts they give their names to farms on which they are situated, the word *cairn* frequently occurring as a prefix to, or as a termination of, names of places and properties. Although the erection of cairns was greatly discouraged on the introduction of Christianity, the practice cannot be said to have quite died out. The valuable ornaments, &c., often found in Scottish cairns is one proof of the esteem in which they were held. A proverbial expression still in use among the Highlanders is ' *Curri mi clach er do cuirn* :' 'I will add a stone to your cairn,' *i.e.* 'I will honour your memory when you are gone.' Many of these monuments may belong to the Later Stone Age, and some have great megalithic chambers and galleries, reminding us of the Uley and West Kennet barrows of the south. A most remarkable group, associated with other primitive monuments, occurs on a small plain near the battlefield of Culloden. Here are large cairns encircled by standing stones at uniform intervals. In the same neighbourhood occur numerous detached monoliths and circular enclosures of small stones. The latter, though much hidden by moss and heather, would appear to indicate the dwellings of the ancient builders of the cairns, who were probably pre-Keltic.

Maeshowe, in Orkney, is a famous example of a long-chambered barrow. Externally it is a truncated cone about 92 feet in diameter, 36 feet in height, and surrounded by a ditch 6 feet deep. The massive stone chambers inside, together with the long passage, make a somewhat cruciform plan. The central chamber was probably 17 feet high, and is connected with three smaller ones of a square shape. The passage is only 4 feet 6 inches high, and probably was closed by a stone door. If, as it seems probable, a race allied to the Lapps were the 'little people' or 'fairies' (see p. 220), they could have walked along the gallery with ease ; but for the tall Kelts who came after them this would have been impossible, they could only have crawled or walked slowly in a very uncomfortable stooping position. Nothing was known of the internal structure of Macshowe until the year 1861, when it was opened in the presence of a select party of antiquaries from Edinburgh. Here, again, tradition comes to our aid ; for it was said to be the abode of a goblin known as the 'Hogboy.' Now the word 'hog' is supposed to be the same as *how* ; in other words, the mound was said to be tenanted by one of the 'how-folk.'

The Norsemen of Orkney had long ago penetrated into this great chambered tumulus. Runic inscriptions were found by the exploring party covering chiefly the walls of the central chamber, but these records, highly interesting and important as they are, do not, as Sir James Fergusson suggests, support the theory that such chambered tumuli were made by the Norsemen, any more than the presence therein of iron weapons does, for such may have been introduced by these men.

In Ireland, where the long barrow is almost unknown, the round barrow, or chambered cairn, prevailed from the earliest pagan period till the introduction of Christianity. These Irish barrows occur in groups in certain localities which seem to have been royal cemeteries. The best known of these was the burial-place of the kings of Tara, situated on the banks of the Boyne above Drogheda, which consists of a group of very large cairns. One of these, at New Grange, is a huge mound of stones and earth, over 300 feet in diameter at the base and 70 feet in height. Around its base are the remains of a circle of large standing stones. The chamber, 20 feet high in the centre, is reached by a passage 70 feet in length.

In the long barrows, and in all the chambered barrows, whether long or circular, the body is usually unburnt and placed in what is known as 'the contracted position,' with the knees drawn up to the head, which is bent forward. Out of a total of 301 burials of unburnt bodies met with by Canon Greenwell in his extensive examinations of barrows on the Yorkshire Wolds, he only met with four instances where the body had been laid out at full length. The former position was evidently the general rule at the time when long barrows were the fashion, that is, in Neolithic times, and perhaps part of the Early Bronze Age. This remarkable position of burial has prevailed in many countries. It is still the custom in parts of Africa, and it was practised by ancient Peruvians, as we can see from some of their mummies, which are tied up in nets in this position. Speaking broadly, it may be said to be very characteristic of the Later Stone Age, or Neolithic times. The Neolithic inhabitants of ancient Egypt had the same custom (*vide* p. 239). This point has given rise to various speculations. In the first place, let us say at once it was certainly *not* due to any desire to compress the body into a small space of grave, for it is found in graves and chambers where there was plenty of room.

Some writers have expressed the opinion that the object of burying a person in this position was to imitate that of a child lying in the womb of its mother, so that the man or woman's entrance into another world should to some extent resemble that of their entrance into this world. The idea may appear rather fanciful, but there is something to be said for it. Another explanation, more generally accepted, is that, this being the position in which most savages rest and sleep, their relations would naturally bury them in that position (in which perhaps they may have died). In Northern Europe the houses, even though partly underground, would not be too warm in winter or summer, and, as their clothing was scanty, they would find such a position both warm and comfortable. This position of rest, as adopted by

P

modern savages, is well illustrated by some large photographs of the people of Borneo in the Ethnological Gallery of the British Museum. We may be allowed to remark that apparently no writer has pointed out how conveniently a corpse so doubled up could be carried, especially if slung in a net or tied up with cords. So perhaps it was more a matter of convenience to the living than of respect to the dead.

Many chambered barrows are known both of the long and the round form in the West of France, and in the Channel Islands, from Brittany to the Gulf of Lyons. They all appear to belong to the Stone Age, and not, as might have been expected from the skill with which they have been constructed, to that of the Bronze. There is in all a most striking absence of that metal, but gold was certainly used even in the Later Stone Age. In Britain and France these monuments appear to have been built by a long-headed, or 'dolichocephalic,' race, whereas in Scandinavia they seem to be monuments of an old round-headed, and probably Turanian race. In a great many cases the body was *not* burned before burial, as we have seen already; but at the same time it would not be wise to lay it down as a rule that in the Stone Age they never practised cremation, while in the succeeding Bronze Age they always did. For example, we find that while in the South-west of England the long-headed race of the long barrows buried without first burning the body, yet, at the same time, on the Yorkshire Wolds cremation was then the rule (to which Canon Greenwell found only one exception).

We have already noticed the absence of metal (except gold). Pottery is seldom found, and what has been found is of a dark colour and often quite plain, or unornamented. It is also of coarser make than the Bronze Age pottery of the round barrows.[1]

---

[1] The reader should not fail to inspect the fine collection of Canon Greenwell in the British Museum Ethnological Gallery, from which he will obtain a good idea of the relics found in British barrows.

# ABODES OF THE LIVING AND THE DEAD

The round barrows belong to a later round-headed Keltic race having a knowledge of bronze. But it should be pointed out that our remarks on the long barrows are confined to 'primary interments,' in which case the skeleton is always a long-headed one and without objects of bronze. But it frequently happened that an old long barrow was used for interments by the race of round-headed people that came after. Such interments are said to be 'secondary,' and they are never at the higher end of the barrow, and are often in the upper layers. Some even belong to the Saxon period, and contain implements of iron, in which case the body is in the extended position. The reader will find it easy to remember that long barrows and long skulls go together, as do round barrows and round skulls (except in Scandinavia). 'The contrast in form,' says Dr. Thurnam, 'between the long skulls from the long barrows and the short or round skulls, which, to say the least, prevail in our Wiltshire circular barrows, is most interesting and remarkable, and suggests an essential distinction of race in the peoples by whom the two forms of tumuli were respectively constructed.'

In long barrows the human bones are often disjointed and lying apart from each other, as if the bodies had been dismembered and the flesh removed before they were placed in the mound. These mounds, then, were not in every case the first place of burial. For various reasons the body appears to have been first placed in some other grave. Both here and in the Scandinavian passage-graves piles of human bones are found, showing that, in some cases at least, they were 'ossuaries,' or places where bones were collected together. This is a well-known practice with various nations (see Chapter XI.).

In many skulls from Wiltshire barrows Dr. Thurnam noticed signs of violent fracture, and came to the conclusion that on the occasion of a chief's funeral slaves were sacrificed (see p. 198) and possibly eaten. There certainly were funeral feasts in connection with interments in long barrows, as is proved by the abundance of animal remains, such as stags' horns and bones of oxen (*Bos longifrons*), wild boar (*Sus scrofa*), &c.

In answer to this, Canon Greenwell, although he was at first inclined to accept the theory, says :[1] 'People were slain in battle, or in consequence of a private feud or quarrel, then as now, and these infrequent fractured skulls may well be the result of such accidents. Had the long barrows, as a rule, contained one or more complete skeletons, surrounded by or associated with others which showed evidence of having been those of persons killed by violence, and broken up as if for use at a feast, then we might have concluded with some probability that it was the habit of these people to immolate, for one purpose or another, certain persons at the time of a funeral. But no such appearances present themselves.' He thinks some of the fractured surfaces of the skulls show signs of fire, and that the fractures themselves may have been caused by the pressure of earth and sods of grass upon bones which had undergone the action of fire whilst covered up.

In the age of Bronze, cremation was the rule, and hence we find the barrow becoming much smaller, round, and containing no stone chambers. It loses also its frequent circle or fence of protecting stones. In fact, it has become quite degraded. 'The round barrows, whether simply conoid or bell-shaped, or of the more elaborate bell or disc forms, are very much more numerous than the long barrows of the same district (Wilts). They much more frequently cover interments after cremation than by simple inhumation, in the proportion of at least three of the former to one of the latter. As, however, the objects found with the burnt bones and with entire skeletons in this class of barrows do not differ in character, but, in addition to implements and weapons of stone (including beautifully barbed arrow-heads of flint), not unfrequently comprise other implements of *bronze*, and also the finer and more decorated sorts of ancient British *fictilia*—

[1] *British Barrows*, p. 545.

Plate VIII.

THE WARRIOR'S COURTSHIP, DENMARK

*All the clothing, weapons, ornaments, taken from actual discoveries in the peat of Denmark*

BRONZE AGE

the so-called "drinking cups" and "incense cups"—we may safely conclude that all are of the same Bronze Age, during which, in this part of Britain, cremation, though not the exclusive, was the prevailing mode of interment.'[1]

[1] See *Some Account of the Blackmore Museum* (Salisbury), published by the Wilts Archæological Society, part i. p. 38.

## CHAPTER X

### THE 'LITTLE-FOLK,' OR FAIRIES AND MERMEN

#### The fairy tales of Science.

SINCE the late Poet Laureate wrote 'Locksley Hall,' the above phrase has become 'current coin,' but in the meantime we have all learned that there is also a Science of Fairy Tales, and students of mythology and folk-lore have opened out new fields of research.  Nor have the archæologists been mere on-lookers, for they also have done not a little to show that many curious tales about fairies, or 'little-folk,' which formerly were looked on as mere inventions of the imagination, are based to some extent upon actual facts from which there is no getting away.

The people who wrote the Scandinavian 'Sagas' were evidently living in the 'Iron Age,' but it is pretty clear that they were still acquainted with an older and more primitive people who had weapons of stone or bone, but not of metal. In a previous chapter (p. 80) we gave Tacitus' description of them.  These Finns, or Fenni, were Laplanders, and may be regarded as survivals, even 1,800 years ago, from the Stone Age.  There is one passage in the 'Sagas' which clearly proves that the arrows of this dwarfish people were of *stone*.  It occurs in a very interesting romance called 'Örvar Odds' Saga.'  A Viking of that name has some magical arrows of stone given to him by an old dwarf whom he met in a forest in Huneland, and the dwarf's prophecy that they would prove serviceable to him came true, for on account of their magical properties he was able to kill with them an old witch who had caused him great loss of men in a battle.

Superstitious notions of the same kind formerly prevailed also among the peasants of Ireland and Scotland.

Mr. E. Lloyd relates [1] that during his journey in Scotland he was much amused with the many different kinds of amulets preserved by the inhabitants. Among these he mentions stone arrows, which were believed by them to have belonged to the elves. In 'Nenia Britannica' (London, 1793, p. 154) is given the figure of a stone arrow-head from Ireland, mounted in silver, and the author states that the peasants call these flint-arrows 'elf-arrows;' that they mount them in silver, and wear them round the throat as amulets, or charms, against 'elf shots' (see British Museum Collection). In this same way Scandinavian peasants wore stone arrow-heads as charms against 'Lapp shots' or Lapp arrows, on the old principle that 'like cures like.' Now this old dwarfish race formerly spread over all Sweden and Denmark. This is proved both by the presence of Lapland skulls in ancient tombs, and also by the fact that some names of places contain words from the Lapp language. Denmark also *may* have been inhabited by this race before the Goths came. To give only two examples, the Lapp words *stock* (sound or inlet) and *garn* (lake) seem to enter into many Swedish local names.

The stories about *dwarfs, goblins, elves,* and *cavern-people* contained in the 'Sagas' cannot be mere inventions of the imagination. They are closely connected also, as we shall see, with green mounds. The dwarfs had been expelled by the stronger and taller Gothic races acquainted with metals, just as in North America the Esquimaux were driven north by the Indians.[2]

There are students of folk-lore and mythology who do not hesitate to say that the dwarfs of the old 'Sagas' were purely

---

[1] *Observations on Wales.*

[2] For a fuller discussion of the subject see Nilsson's *Primitive Inhabitants Scandinavia.* We can only give here a few of the arguments of this high authority.

mythical, and only meant to typify certain powers of Nature. But the descriptions given of these people are confirmed by archæological discoveries; they are also too minute and matter-of-fact to be pure inventions. Of course, the dwarfs are given supernatural powers, but that is only what might be expected. Poets in those days adorned their stories with extravagances. It was only a little 'poetic licence!' Homer did the same, and yet there is probably a good deal of historical truth underlying the 'Odyssey' and the 'Iliad.' When one race described another with which it had come in contact, there was usually much exaggeration.

Thus the Esquimaux of North America have described the English people to travellers in terms quite as fanciful and extravagant. They firmly believed that white men were giants; that they had wings; that they could kill with a glance of the eye, and swallow a whole beaver at a mouthful! So we need not be at all surprised when we read in the 'Saga' of Olaf Trygvadson about a couple of Finns or Laplanders with whom the fair Gunhild was staying in order to learn the science of sorcery; that 'they also could kill with a glance, because when anything living encountered their eye, it fell down dead at once, and when they were angry the earth recoiled at a look. They missed nothing at which they aimed; they could follow the trail like dogs, on frozen as well as on damp ground, and they could run in snow-shoes so swiftly that neither man nor beast could overtake them.'[1]

In some sagas dwarfs are mentioned as living in mountain caves; in others, their dwelling is said to be underground. P. Læstadius says: 'There is a saga which tells us how some hostile people once discovered such an earth-cavern by hearing a woman from within calling out to somebody who was in an inner room to fetch the cooking-ladle. This was overheard by the enemy outside, who forthwith broke in upon them, and slew those who were in the cavern.' This evidently refers to a dwelling similar to the gamme of the Laplander, described in last chapter,

[1] Nilsson, *loc. cit.*, p. 209.

and in *one* saga a dwarf is expressly mentioned as living in a gamme.

These primitive people, living in scattered villages, went on using their stone weapons, as of old, long after their conquest by Gothic peoples, as the following Scandinavian story shows. A peasant who had gone out to look for his horses wandered about nearly the whole day without finding them. Towards evening, when he came into a previously unfrequented track, he met with a dwarf who was working in the forest. The dwarf, on perceiving the peasant close beside him, became so alarmed that he immediately threw down his tools and ran away as fast as he could. The peasant then approached the place where the dwarf had been, and found there an axe, a chisel, and some other tools; but he could not make any use of them, because the dwarf, before running away, had transformed them all into stone!

Of course, the real meaning of this last passage is that the dwarf's tools were made of stone instead of metal. But the story is highly useful, for it shows that dwarfs worked like other people, that they were afraid of their conquerors, that they were supposed to be skilled in magic or sorcery. These mysterious little people were evidently dreadfully thievish, and so cleverly did they steal things and make themselves scarce, that the popular belief was they could render themselves invisible. In Swedish and Danish folk-sagas there are stories of how the goblins, or dwarfs, attended a wedding, but invisibly, and ate up all the food of the guests. In these prosaic days such incidents are usually termed thefts. Among the mountains of Northern Norway, where Laplanders wander with their herds of reindeer, these people apparently are still adepts in the old art.

Occasionally marriages took place between members of the two races; but only occasionally, for dwarf women were seldom good-looking, and two races so different in every way would not be on too friendly terms. From the 'Sagas' we learn that they were ugly, and lived in mountain caverns, hillocks, or earthen mounds, mostly in solitary tracks; they had children and sometimes servants, and were believed to have hoarded up much silver and copper! Sometimes they wanted to borrow some things, and then they approached the houses of the country people in the evening to ask for them. They never dared to pass the threshold, but stood outside the house, calling in a loud voice for what they wanted. Generally they sent one or two of their children on such an errand. If what they asked for was given to them, it was always found lying early in the morning, a few days after, in the same place, and beside it, as payment, a silver coin, or something else of value. So they had feasts of their own. The hatred between the two races is revealed in the older sagas, which tell how the poor dwarfs were persecuted, shot through with red-hot arrows, or cut to pieces with axes. These also describe them as a degraded race, often thievish, often generous, but with whom nobody wished to be closely united. They were also said to be cowardly. Laplanders wear a grey kirtle of reindeer-skin and a blue or red cap, and the dwarfs were said to wear similar garments.

Dwarfs and giants seem to go together naturally, and Professor Nilsson believes that all the ancient sagas about jotnar, or giants, originally emanated from the dwarf people, to whom doubtless the Goths would appear as giants.

It would be easy to cite many examples from history and tradition to show that when a small race encountered one more powerfully built, fabulous legends arose from an excited imagination. In a rude and timorous people a foolish panic might thus be easily created. When Moses sent spies from the desert of Paran in order to glean some information about the ancient people of Canaan, his spies returned, saying that the Anakim, or children of Anak, were living there, adding: 'We were in our own sight as grasshoppers, and so we were in their sight.' When the congregation heard this they 'lifted up their voice and cried, and the people wept that night.' So frightened were they by these reports that at first they even wished to return to the bondage of Egypt. 'And the land of the children of Ammon also was accounted a land of giants ; giants dwelt there in old times.'[1]

Again, the Germani, or ancient Germans, were a tall race, and when Julius Cæsar arrived at Besançon, a report was spread by the Gauls and merchants throughout the Roman army of the gigantic stature of the Germani, with the result that the former were panic-stricken. Some of the officers went home, others 'wept and groaned in their tents.'[2]

Let us now return to our fairies, or little-folk, tracing the evidence of their former presence in Scotland and Ireland, evidence of the same kind, derived from tradition, from archæological discovery, from skulls, and from mythology and philology. Just to give one example from the latter before passing on. Has the reader ever considered what is the meaning of the word 'fox-glove' applied to the beautiful tall wild-flower of that name ? It is said by some people to be an interesting survival from the time when people believed in fairy folk, for the pretty and slender red flowers were to them the 'folk's-gloves,' now shortened into fox-glove ! When all that is magical and miraculous, or superstitious, is taken away, there yet remains in folk-lore and fairy tales a certain residuum of truth. This has already been amply proved by archæological and other researches. How delighted must Mr. Ruskin and all true followers of our great teacher and prophet be to learn that, after all, we shall not have to give up our fairies ! And what an anti-climax must such a result appear to those hard, unsentimental scientific workers and thinkers who were wont to consider fairy tales as nothing but pure 'stuff and nonsense'!' It must be somewhat humiliating to such—if there are any left—to reflect that they must no longer dare to despise fairies, but are compelled, in the sacred name of Science (with a *very* big S), to pay homage to them !

Only last year the writer was once more impressed with the

[1] Numbers xiii. 33 ; xiv. 1, 2. See also Deut. ii. 10, 20, 21.
[2] Cæsar, *De Bello Gall.*, i. 39.

importance and influence of fairies in witnessing a magnificent performance of Wagner's sublime opera *Tannhaüser*, at Drury Lane Theatre. The 'Venus-berg' was rightly represented by the scenic artist as an underground palace of the Queen of the Fairies in a great berg, or green hill, into which the hero is tempted by the attractions of the dwarf-women. And how wonderfully has the great composer expressed in music, which is far above *words*, the feelings of horror and hatred with which the dwarfs are regarded by their conquerors, who, of course, consider them as devils in league with the powers of darkness !

Again, have we not in Browning's 'Pied Piper of Hamelin' another example of the magical and thievish arts attributed to the dwarf people? For did not the said Piper get by a clever stratagem his golden ducats from the town councillors, and, having done so, did he not charm the rats that so infested the town into the river, and all the children who followed him into a great cavern in a mountain? Allowing for 'poetic licence,' we at once see in the mountain and its fairy cavern, or palace, simply a big green hillock with its underground passages and chambers !

It is not necessary to go as far East as Mykene to find the chambered mound with its dry-stone walls and 'pelasgic' arch, for many a seeming hillock may also be a 'treasure house.' Mediæval castles were often built on such mounds. Thus, according to local tradition, the hill upon which Kenilworth Castle is built was once inhabited by fairies, to whom are attributed the same characteristics as elsewhere.

The 'Castle Hill' of Clunie, in Perthshire, is another example. It is a large green mound, partly natural and partly artificial, on the top of which are the ruins of a very old building. A hundred years ago there were old people living near who said that they had seen an opening in the mound which led to subterranean chambers. More than one 'fairy-knowe' of the present day, not yet explored, has a small hole on the top, in which, when a stone is dropped, a rumbling noise is heard as the stone rolls along some underground chamber. According to Sir Walter Scott, the peasants are well aware of these underground places. He says : ' Wells, or pits, on the top of hills were supposed to lead to the subterranean habitations of the fairies.' In not a few legends it is said that men descending such pits engaged in hand-to-hand fights with the dwellers in these abodes.

But sometimes the dwarfs themselves began the attack. Even heraldry bears its testimony to the former existence of these people, for 'a savage issuing from a mount' was once a well-known bearing in Scottish heraldry.[1]

Mr. J. F. Campbell records a Ross-shire tradition of a dwarf who inhabited 'The Tawny Hill of Gairloch,' and who was the terror of the neighbourhood. Before he was himself slain he killed many of the taller race, none of whom dared to venture near his hillock after dusk. He was at length killed by a local champion, who was celebrated as a slayer of dwarfs. The story says that Uistean (the champion) climbed to the top of the hillock and attacked the dwarf, who emerged from its ' pit.'

In the valley of the Boyne, near Doune, in Ireland, is a very large mound, or knowe, called the Brugh of Boyne. It is even larger than the Maes-howe (Orkney). In this underground palace a certain Angus Og 'magnificently dwelt.'

This person is said to have been the king of the Tuatha De Danann. They are sometimes spoken of as 'the Dananns,' sometimes as 'the Tuatha De, or Dea.' The word *Tuatha* means 'people.' It is said they came from ' Lochlin ' (Scandinavia, or North Germany). Then they crossed to Ireland. Two centuries later the Gaels (or Milesians) came to Ireland. It

---

[1] **A** good many writers have incidentally borne their testimony to these traditions. But we are chiefly indebted for the information here given to Mr. David MacRitchie, whose works on *The Testimony of Tradition*, and *Finns, Fairies, and Picts* (Kegan Paul, 1890 and 1893), show a great deal of careful research.

was at this time that Angus was king of the Tuatha De Danann. The Gaels successfully invaded Ireland, but the former people could not make satisfactory terms with the victors. And so it was agreed that the matter should be laid before the first person whom a party of deputies from either side should happen to encounter at the outskirts of a certain town on an appointed day, and the man's decision should be final. On this occasion the druid or wizard of the Gaels was more than a match for the dwarfs, clever as they were at magic; for it was arranged between this man and his party that the first person whom the deputies should meet was to be the druid himself. The poor Dananns were fairly done; for the first man the delegates met was (apparently) a strolling harper. 'It is a great thing thou hast to do to-day, good master of the sciences!' was the greeting of Angus. 'What have I to be doing to-day,' quoth the wise man, 'except to go about with my harp and learn who shall best reward me for my music?' 'Thy task is far greater than that,' answered Angus; 'thou hast to divide Ireland into two equal portions.' Thereupon the *druidh* (druid), having obtained the promise of either side that they would abide by his decision, pronounced as follows :—

'This then is my decision. As ye, O magical Dananns, have for a long period possessed that half of Ireland which is above ground, henceforth the half which is underneath the surface shall be yours, and the half above ground shall belong to the sons of Miledh (the Milesians). To thee, O Angus, son of the Dagda, as thou art king of the Tuatha De Danann, I assign the best earth-house in Ireland, the white-topped *brugh* of the Boyne. As for the rest, each one can select an earth-house for himself.' Against this grotesque decision there was no appeal, and the poor dwarfs surrendered the surface of Ireland to the Gaels, 'retaining only the green mounds known by the name of Sidhe, and then being made invisible by their enchantments became the Fir Sidhe, or Fairies of Ireland.'[1]

[1] Skene's *Celtic Scotland*, i. 178 and 220, iii. 106.

This tradition shows that the Tuatha De Danann were themselves mound-dwellers, and the Gaels made them keep to their own underground habitations, with perhaps a small reservation of territory, just as in North America the white men have allowed the Indians to live in reserves. The Tuatha De Danann then were the Sidhfir, or Fairies, of Irish tradition. Brugh is obviously the same word as berg, or burgh, and has survived in place names such as Edinburgh, Roxburgh. The same root may be traced in the old Greek Pergammon.

But the fairies appear to change their *rôle* pretty frequently: sometimes they are malicious thieves, at others sorcerers; sometimes hated, yet again sometimes the objects of adoration. We all remember how in fairy tales the fairies must be propitiated. The biographer of St. Patrick says of him:

> He preached threescore years
> The Cross of Christ to the *Tuatha* [people] of Feni,
> On the *Tuatha* of Erin there was darkness.
> The Tuatha adored the Side.

The 'little people' maintained their exalted character long after they were conquered. Even to this day the common people of Ireland speak of the inhabitants of the brughs as 'the gentry.' A Gaelic poem of the fifteenth century says:

> Thou, the son of noble Sabia,
>   Thou the most beauteous apple rod;
> *What god from the Bru of the Boyne*
>   Created thee with her in secret?

Probably some of the gods of this race were incorporated into the mythology of the Keltic people, and doubtless much of the superstition of the bronze-using people was also derived from the Later Stone Age. The Russians also believe in 'Tshuds,' or vanished supernatural inhabitants of their land. With this we may compare the Egyptian tradition that before Mena, the first king of the first dynasty, the gods reigned in Egypt.

Referring to the island of Sylt, off the Schleswig coast, Mr. William George Black says, with regard to a story of 'Finn, the king of the dwarfs:' 'These were an odd, small, tricky people whom the Frisians found in Sylt when they took possession. They lived underground, wore red caps, and lived on berries and mussels, fish and birds and wild eggs. They had stone axes and knives, and made pots of clay. They sang and danced by moonlight on the mounds of the plain which were their homes, worked little, were deceitful, and loved to steal children and pretty women; the children they exchanged for their own, women they kept. Those who lived in the bushes, and later in the Frieslanders' own houses, like our own brownies, were called "Pucks," and a sandy dell near Braderup is still known as Pukthal. . . . They had a language of their own, which lingers yet in proverbs and children's games. The story of King Finn's subjects is evidently one of those valuable legends which illuminate dark pages of history. It clearly bears testimony to the same small race having inhabited Friesland in times which we trace in the caves of the Neolithic Age, and of which the Esquimaux are the only survivors.' Mr. Black visited one of those green mounds said to have been inhabited by this Finn from Sylt, and he states that when it was first scientifically examined in 1868 it was found to contain remains of a fireplace, bones of a small man, some clay urns, and stone weapons.

The late Mr. J. J. Campbell, of Islay, says: 'I believe there was once a small race of people in these islands, who are remembered as fairies, for the fairy belief is not confined to the Highlanders of Scotland. . . .' This class of stories is so widespread, so matter of fact, hangs so well together, and is so implicitly believed all over the United Kingdom, that I am persuaded of the former existence of a race of men in these islands who were smaller in stature than the Kelts; who used stone arrows, lived in conical mounds like the Lapps, knew some mechanical arts, pilfered goods and stole children, and were perhaps contemporary with some species of wild cattle and horses and great auks, which frequented marshy ground, and are now remembered as water-bulls and water-horses, and boobries, and such-like impossible creatures.

He suggests that certain 'fairy herds' in Sutherlandshire were probably reindeer; that the 'fairies' who milked those reindeer were probably of the same race as the Lapps, and that not unlikely they were the people historically known as Picts.

According to Mr. MacRitchie, 'Santa Claus,' another cherished fiction of childhood, turns out to be a good-natured Lapp, fond of children. Referring to a German illustration of this highly popular and good fairy, he says : 'The German idea, then, of this good magician is that he is a thick-set, bearded little man, whose heavy fur denotes that his home lies in the North, and whose reindeer team, harnessed to the sledge in which he has travelled, indicates that, like the Lapp and the Aino [of Japan], he not only lives in a country where reindeer abound, but he has learned to tame them and make them serve his purpose.'

The well-known story of Child Roland, made familiar to many by the words 'Childe Roland to the dark tower came . . .'[1] is probably another example of a popular tradition about the dwarfs and their habit of stealing young women. 'Certainly,' says Mr. Joseph Jacobs,[2] 'the description of the dark tower of the King of Elfland, in "Child Roland," has a remarkable resemblance to the dwellings of the "good folk" which recent excavations have revealed.' Roland is seeking his lost sister. Henwife tells him to go on a little further till he comes to a round green hill surrounded with terrace rings from the bottom to the top, and to go round it three times in a direction opposite to the sun's course (*i.e.* 'Widershins') and say each time, 'Open door; open door.' Now it is important to note that Mr. G. L. Gomme, in one of his

---

[1] Shakespeare's *King Lear.*
[2] See his *English Fairy Tales* (with exquisite illustrations by Mr. Batten); notes, p. 238.

works,[1] has given good reasons for believing that the old pre-Keltic and non-Aryan inhabitants of Great Britain practised terrace cultivation along the sides of the hills. Such terraces the writer has seen in Cambridgeshire, Wilts, and Dorset, and has often tried to convince people living in the country that they are *not* natural, for no geological action could possibly have formed them. In Wilts, at least, there are no traditions about them. One reason for this terrace cultivation probably was the fact that in those old days, perhaps 3,000 years ago, the valleys were so densely wooded as to be a great hindrance—especially to a people with only *stone* weapons.

In this old legend we have evidence of the association of the King of Elfland with the cultivation of the ground by means of terraces. But to return to Child Roland, it is very likely that we have here an idealised picture of a 'marriage by capture' of one of the diminutive non-Aryan dwellers of the green hills with an Aryan maiden, and her recapture by her brothers. Mr. Jacobs shows that Milton got hold of the same story and used it in his 'Comus.'

Mr. MacRitchie further believes that the Picts (or Pechts) were of this non-Aryan race, but whether that conclusion will ever be established it is very difficult to say at present. Other writers believe them to be Keltic. The popular traditions about these people ascribe to them a low stature, but superhuman strength—'unco' wee bodies, but terrible strang.' The late Mr. Robert Chambers, in putting together the popular Scotch beliefs regarding these people, not only states that they were 'short wee men,' but, he adds, 'the Pechts were great builders ; they built a' the auld castles in the kintry ; and do ye ken the way they built them ; I'll tell ye : They stood all in a row from the quarry to the place where they were building, and ilk ane handed forward the stanes to his neebor, till the hale was biggit [builded].' The round tower of Abernethy is said to have been built by Picts.

[1] See *The Village Community* (Walter Scott, 1890).

Earth-houses, Picts-houses, or weems, are very abundant in Scotland in many places, especially on the upper reaches of the Don, in Aberdeenshire. In the low country they are called 'erd-houses,' and are there said to be the hiding-places of the aborigines. So numerous are they in some places, that they may be said to form subterranean villages, the fields being literally honeycombed with them ; but they are not easy to find. Sometimes one is to be found under an unploughed patch in a field, with a few stones above ground. At other times, a man finds a hole between two projecting stones, and on letting himself down through this he finds an underground gallery leading to chambers. The masonry is of that simple kind, 'Cyclopean,' with no mortar, no carvings, and no inscriptions, and no marks of tools. The absence of all these things is negative evidence of a very important kind which must be borne in mind. Uncouth, gloomy, and unadorned as they are, yet a wonderful amount of labour and mechanical skill must have been devoted to them.

To the ordinary observer the level heath, or moor, or hillside under which they lie presents no appearance of having ever been disturbed, so that he would never suspect that he is walking over a house of some distant prehistoric period.

It may, perhaps, not be out of place to remark here that Tacitus, in writing of the customs of the Germans, says : 'They dig caves in the earth, where they lay up their grain and live in winter. Into these they also retire from their enemies, who plunder the open country, but cannot discover these subterranean recesses.' No doubt one of the chief objects for which they were built was concealment ; but at the same time warmth and protection against a cold and inclement climate must also have been powerful motives.

In the sagas of the Norsemen there are several references to underground houses. They vary greatly in size, form, and arrangement. Here is a description of one which the writer has visited at Kingussie (Inverness-shire), called the 'Cave of

Raitts,' on the estate of Mr. Macpherson, of Belleville, and not far from the house of that name.

It is an erd-house, the only one of this class of antiquarian remains that exists in Badenoch. It is in the form of a horse-shoe, which has one limb truncated, about seventy feet long, eight feet broad, and seven feet high. The walls gradually contract as they rise, and the roofing is formed by large slabs thrown over the approaching walls. Part of the roof has already fallen in. The common tradition is that it was inhabited by a band of savage robbers, said to have been a remnant of the barbarous tribes who, after the overthrow of the Comyns in the district, infested the wilds of Badenoch and plundered the peaceable inhabitants. But at last the whole gang were put to death.

An Irish chronicler (of about the twelfth century) says that such places were plundered in Ireland by Danes of the ninth century. The following story, taken from the tenth-century saga of Thorgils,[1] is interesting, as showing that a leafless tree, or branch, stuck in the ground was taken for a sign that an underground house existed there. This Thorgils and another pirate, Gyrd, had joined together on a plundering expedition.

'Now they harried during summer with much gain, and exterminated many robbers and evil-doers, but leaving genuine farmers and traders in peace. Towards summer they came to Ireland [to a place] where in front of them they discovered a forest. Just after entering the forest they came to a spot where they saw a tree whose leaves had fallen off. They pulled up the tree [evidently a sapling], and beneath it they found an underground chamber, wherein they saw men with weapons. Thorgils proposed to his people that whoever should be the first to go into the earth-house should become entitled to the three objects of booty which he desired, to which all agreed except Gyrd. Then Thorgils sprang down into the chamber, and encountered no opposition; and there were two women there, one of whom was young and

---

[1] MacRitchie, *Underground Life*, p. 28.

beautiful, and the other old, yet not without good looks. Thorgils went about the chamber, whose roof rested upon upward-bent beams; he had a mace in his hand, wherewith he smote about him on either side, so that all fled before him. Thorstein went with him, and then they came out of the earth-house, and took the women, the young one as well as the elder, with them to the ships. The people of the place now set out in pursuit of them, and Thorgils getting on board, they steered out from the shore. Now a man of the host which was pursuing them stepped forward and harangued them, but they understood not his speech. Then the captured women interpreted his story to them in Norse, and said: " He will resign his claim to the goods you have taken, if only you will let us go. This man is an earl, and my son; but my mother's kindred are from Vik, in Norway. Follow my counsel, then will you best derive benefit from this rich booty, for trouble comes with the sword. My son is named Hugh, and he has proffered to thee, O Thorgils, other goods, rather than that you should carry me away, which could not be of any profit to you." Thorgils agrees to their request, and brings them to land. The earl went joyfully towards Thorgils, and presented him with a gold ring; his mother gave him another, and the maiden gave him a third. Thereafter they bade each other a friendly farewell.'

But, if some tribes of the 'little-folk' lived underground, like the older race in Mr. Wells's clever story, 'The Time Machine,' it appears to be equally true that others hailing from Norway spent much of their time on the sea, in little skin-boats or kayaks. So skilful was their management of the frail craft, and so rapid their rate of travel, that they made journeys across the North Sea from Norway to Shetland and other parts of the British Isles. Hence it seems very probable, as certain writers believe, that these little people, being so much at home in the water, became the mermen and mermaids, or merwomen, of folk-lore. If this is so, and the idea appears to work out very well, we have yet another illustration of what was said at the beginning of this chapter about the

Science of Fairy Tales. Fairies have been 'run to earth' and traced to their subterranean haunts ; and now we shall endeavour to show that mermaids and such folk are no mere mythical fancies, and can be run not to *earth* but to *water !*

We are no longer in bondage to the now old teaching of not many years ago, that everything that was extravagant, miraculous, or incredible in mythology and folk-lore was pure invention, and, moreover, invented from nothing ! But there is an old saying, *Ex nihilo nihil fit* (nothing comes from nothing), and the subject now before us is a very good example to show that most of the marvellous incidents and stories of ancient mythology have a certain element of truth, and are founded, though with much exaggeration and alteration, on actual facts. The 'Mythical School' and all its ideas is now superseded by one more scientific in its method and based on anthropological studies. The former was 'made in Germany !'

We shall now lay before the reader in brief form some of the facts and arguments of certain writers [1] who maintain that the mermen were a Finnish people allied to the Laplanders.

Dr. Karl Blind, in one of the papers referred to in our foot-note, remarks : 'It is in the Shetland tales that we hear a great deal of creatures partly more than human, partly less so, which appear in the interchangeable shape of men and seals. They are said to have often married ordinary mortals, so that there are, even now, some alleged descendants of them, who look upon themselves as superior to common people. In Shetland, and elsewhere in the North, the sometimes animal-shaped

---

[1] The following account is largely derived from two very interesting books by Mr. David MacRitchie, which we have read with great pleasure, viz. *The Testimony of Tradition* (Kegan Paul, 1890), and *Finns, Fairies, and Picts* (Kegan Paul, 1893). The above author refers to an interesting series of papers by Dr. Karl Blind, on 'Scottish, Shetlandic, and Germanic Water Tales,' contributed to *The Contemporary Review*, 1881, and *The Gentleman's Magazine* of 1882. Mr. G. L. Gomme, in his *Village Community*, a most careful work, speaks of the non-Aryan element in our population.

creatures of this myth, but who in reality are human in a higher sense, are called Finns. Their transfiguration into seals seems to be more a kind of deception they practise, for the males are described as most daring boatmen, with powerful sweep of the oar, who chase foreign vessels on the sea. At the same time they are held to be deeply versed in magic spells and in the healing art, as well as in soothsaying. By means of a "skin" which they possess, the men and the women among them are able to change themselves into seals. But on shore, after having taken off their wrappings, they are, and behave like, real human beings. Anyone who gets hold of their protecting garment has the Finns in his power. Only by means of the skin can they go back to the water. Many a Finn woman has got into the power of a Shetlander and borne children to him; but if a Finn woman succeeded in re-obtaining her sea-skin, or seal-skin, she escaped across the water. Among the older generation in the Northern Isles persons are still sometimes heard of who boast of hailing from Finns; and they attribute to themselves a peculiar luckiness on account of that higher descent.'

The question is, Who are these Finns of the Shetland story? It seems highly probable that they are, when stripped of their magical powers, members of the Ugrian or non-Aryan race dwelling in Finland, in the most northern part of Norway, and in parts of Russia.

Many, or most, naturalists say that the mermaid myth is founded on the seal, with its very human face, and so it may be, or probably is, in part; but these Finns in their seal-skin dress and seal-skin canoe, or kayak, seem to be, as it were, the other half of the myth, and probably are still more necessary for its explanation. They would be the old 'sea-dogs.' Their 'protecting garment' evidently was the kayak; without that they were helpless, for, as was said above by Dr. Karl Blind, 'only by means of the skin can they go back to the water.' The skin-boat was like a part of them.

These Finns were wont to pursue boats at sea, and it was considered dangerous in the extreme to say anything against them. Silver money was thrown overboard to prevent their doing any damage to the boat. In the seal form they were said to come ashore every ninth night to dance on the sands. No doubt, when on shore, and clothed in an ordinary seal-skin *dress* (not the canoe), they may have looked very like seals, or rather appeared to do so to the simple-minded fisher-folk of Shetland. There are traditions of Shetlanders being deserted by their Finnish wives who preferred to return to their Finnish husbands or lovers when they got the chance. The race was said to come from Norway. In a Shetland spell-song, referring to the cure of toothache, the Finn, appearing in the character of a magic medicine-man, hails from that country.

> A Finn came ow'r fa Norraway
> Fir ta pit töth-ache away,

and there is not much doubt that some of them did come from Norway. The speed of the kayak is wonderful, and no water can get into it (unless the skin cracks), as the man and the seal-skin completely fill up the round hole in the centre of it.

Dr. Robert Sinclair, speaking of the capture of Finnish brides by Shetlanders, says : ' Each district almost has its own version of a case where a young Shetlander had married a female Finn. They were generally caught at their toilet in the tide-mark, having doffed the charmed covering [canoe] and being engaged in dressing their flowing locks while the enamoured youth, by some lucky stroke, secured the skin, rendering the owner a captive and victim of his passion. Thus it was that whole families of a mongrel race sprang up, according to tradition. The Finn women were said to make good housewives. Yet there was generally a longing after some previous attachment; if ever a chance occurred of recovering the essential dress, no newly-formed ties of kindred could prevent escape and return to former lovers. This was assiduously guarded against on one side and watched on the other; but, as the story goes, female curiosity and cunning were always more than a match for male care and caution, and the Finn woman always got the slip.'

The poor forsaken husband, on finding the mother gone and the little ones crying, would run with all his might through the standing corn to the shore, but only to see his gude wife in the arms of a merman, or seal (as he might appear to be in his sealskin dress). She cried :

>Blessin's be wi' de,
>  Baith wi' de and da bairns !
>Bit do kens, da first love,
>  Is aye da best !

In one of Matthew Arnold's most delightful poems, 'The Forsaken Merman,' we have evidently an opposite case to this, for there the husband is a merman, and the wife leaves him and his children for a lover on the land. She hears the far-off sound of a bell, and rushes from the sea to the little church on the shore to pray for her kinsfolk. The husband and children go to look for her and beg her to return, but she heeds them not. We cannot refrain from quoting the last verse of this exquisite poem, which we are glad to hear is taught in many of the national schools :

>But children at midnight,
>  When soft the winds blow,
>When clear falls the moonlight,
>  When spring-tides are low ;
>When sweet airs come seaward
>  From heaths starr'd with broom,
>And high rocks throw mildly
>  On the blanch'd sands a gloom :
>Up the still glistening beaches,
>  Up the creeks we will hie ;
>Over banks of bright seaweed
>  The ebb-tide leaves dry.
>We will gaze, from the sand-hills,
>  At the white sleeping town ;
>At the church on the hill-side,
>  And then come back down.

Singing, 'There dwells a loved one,
But cruel is she :
She left lonely for ever
The kings of the sea.'

The following testimony from an old writer [1] is particularly interesting : 'There are frequently Fin-men seen here upon the coasts, as one about a year ago on *Stronsa*, and another within these few months on *Westra*, a gentleman with many others in the isle looking on him nigh to the shore ; but when any endeavour to apprehend them, they flee away most swiftly, which is very strange, that one man sitting in his little boat should come some hundreds of leagues [miles ?] from their own coasts, as they reckon *Finland* to be from Orkney. It may be thought wonderful how they live all that time, and are able to keep the sea so long. His boat is made of seal-skins, or some kind of leather ; he also hath a coat of leather upon him, and he sitteth in the middle of his boat with a little oar in his hand, fishing with his lines. And when in a storm he seeth the high surge of a wave approaching, he hath a way of sinking his boat till the wave pass over, lest thereby he should be overturned. The fishers here observe that these *Finmen*, or *Finlandmen*, by their coming drive away the fishes from the coasts. One of their boats is kept as a rarity in the *Physician's Hall at Edinburgh.*' [2]

It might seem at first sight impossible that these people could have come even from Norway, but a recent traveller has stated that a skilled Esquimaux can go even eighty miles in his kayak in one long day. An inlet at Bergen is called ' Fen's Fiord.' A special caste known as Strils, or Streels, who are very primitive, still inhabit the numerous islands that protect Bergen from the ocean. They speak Norwegian in a way which is difficult to understand, and it is believed that their idioms come from the land of the Samoyedes.

[1] Brand's *Brief Description of Orkney, Zetland, &c.* (1701).
[2] This boat *may* be in the Edinburgh Museum of Science and Art.

TWO BRITISH WARRIORS
BRONZE AGE

Dr. Robert Sinclair also says : 'Here, then, the Finns are men of human origin; remaining intelligent men in their sea-dog raiment; coming from Norway, not swimming like marine animals, but rowing between Shetland and Norway, namely, to the town of Bergen, which lies in the southern . . . part of Norway. As strong men at sea, they row with magic quickness. . . . Each one of them . . . must have his specially prepared skin. . . . There is nothing here of the swimming and dipping down of a seal.'

But, of course, the Finnish dwarfs, with their little round skulls, were not the only inhabitants of Europe prior to the Aryan invasion. There was evidently a long-headed race which, as we showed in the last chapter, were probably buried in some of the long barrows.

Representatives of an old pre-Keltic race are now chiefly found in the Basque Provinces of Spain, in the West of Ireland, and some parts of Wales and the Highlands of Scotland. They are generally known as Iberians, but also by the terms Silurian, Euskarian, Basque, and Berber (the Berbers may be related to them). Though now cut up into isolated fragments by many invasions and settlements of the Aryan race, archæologists have traced them by means of their tombs and cave burials, which show them to have been in possession of the whole of Europe north and west of the Rhine during the Later Stone Age. To this day they are represented by the dark and short Welshman or Highlander occasionally met with by the traveller. The late Mr. Campbell, of Islay, thus describes one : 'Behind the fire sat a girl with one of those strange foreign faces which are occasionally to be seen in the Western Isles, a face which reminded me of the Nineveh sculptures, and of faces to be seen in St. Sebastian. Her hair was as black as night, and her clear dark eyes glittered through the peat-smoke. Her complexion was dark, and her features so unlike those who sat about her, that I asked if she were a native of the island, and learned that she was a Highland girl.'

This race probably is that of the dolmen-builders, for they appear to be connected with the dolmens, or tombs made of a few big stones placed together; but as we shall have something to say about these curious structures in the next chapter, we will content ourselves with quoting the following description of Neolithic times by Professor Boyd Dawkins :[1]

'If we could in imagination take our stand on the summit of a hill commanding an extensive view in almost any part of Britain or Ireland in the Neolithic period, we should look upon a landscape somewhat of this kind. Thin lines of smoke rising from among the trees of the dense virgin forest at our feet would mark the position of the Neolithic homesteads, and of the neighbouring stockaded camp which afforded refuge in time of need; while here and there a gleam of gold would show the small patch of ripening wheat. We enter a track in the forest, and thread our way to one of the clusters of homesteads, passing herds of goats and flocks of horned sheep, or disturbing a troop of horses, or small short-horned oxen, or stumbling upon a swineherd tending the hogs in their search after roots. We should probably have to defend ourselves against the attack of some large dogs, used as guardians of the flock against bears, wolves, and foxes, and for hunting the wild animals. At last, on emerging into the clearing, we should see a little plot of flax, or small-eared wheat, and near the homestead the inhabitants, clad some in linen and others in skins, and ornamented with necklaces and pendants of stone, bone, or pottery, carrying on their daily occupations. Some are cutting wood with stone axes with a wonderfully sharp edge, fixed in wooden handles, with stone adzes and gauges, or with little saws composed of carefully notched pieces

[1] *Early Man in Britain*, p. 271.

of flint about three or four inches long, splitting it with stone wedges, scraping it with flint flakes. Some are at work preparing handles for the spears, shafts for the arrows, and wood for the bows, or for the broad paddles used for propelling the canoes. Others are busy grinding and sharpening the various stone tools, scraping skins with implements ground to a circular edge, or carving various implements out of bone and antler with sharp splinters of flint; while the women are preparing the meal with pestles and mortars and grain-rubbers, and cooking it on the fire, generally outside the house, or spinning thread with spindle and distaff, or weaving it with a rude loom. We might also have seen them at work at the moulding of rude cups and vessels out of clay which had been carefully prepared.'

It is unfortunate that so little evidence remains of the kind of clothing worn by the men of the Later Stone Age in Europe. In North America the age of metals came very much later, for many tribes of Indians were practically living in the Stone Age even so late as the Spanish conquest! In a Spanish work alluded to by Professor Boyd Dawkins,[1] a most interesting account is given of the prehistoric antiquities of Andalusia. The author, Don M. Gongora y Martinez, describes several interments in the Cueva de los Murciélagos, a cave running into the limestone rock, out of which the grand scenery of the southern part of the Sierra Nevada has been to a great extent carved. In one spot, a group of three skeletons was met with, one of which was adorned with a plain coronet of gold, and clad in a tunic made of esparto grass, finely plaited, so as to form a pattern which resembles some of the designs on gold ornaments from Etruscan tombs. Further on a second group of twelve skeletons was found lying in a semicircle, around one considered by Don Manuel to have belonged to a woman, covered with a tunic of skin and wearing a necklace of esparto grass, a marine shell pierced for suspension, the carved tusk of a wild boar, and earrings of black

[1] *Cave-hunting*, p. 209.

stone. A variety of articles, such as baskets and sandals, flint flakes, axes, awls, a wooden spoon, &c., were found here also. But there is no positive proof that this burial took place in the Stone Age.

It is probable that the earliest traditions of old Egypt about the origins of their religion and arts point to a time which may be considered as representing the transition from the age of Stone to that of Bronze (but this is only a speculation of the writer's).

The ancient Egyptians attributed the beginnings of civilisation to their favourite god, Osiris, who once reigned on earth as 'the good god,' and afterwards became their judge in the next world. He married his sister Isis, who became queen-regnant with him. The Egyptians were as yet but half-civilised; they were cannibals, according to Maspero, and, though occasionally they lived upon the fruits of the earth, they did not know how to cultivate them. Osiris taught them the art of making agricultural implements— the plough and the hoe, field labour, the rotation of crops, the harvesting of wheat and barley, and vine culture. Isis weaned them from cannibalism, healed their diseases by means of medicine or of magic, united women to men in legitimate marriage, and showed them how to grind grain between two flat stones, and to prepare bread for the household. She invented the loom with the help of her sister Nephthys, and was the first to weave and bleach linen. There was no worship of the gods before Osiris established it, appointed the offerings, regulated the order of ceremonies, and composed the texts and melodies of the liturgies. He built cities, among them Thebes itself, according to some, though others declare that he was born there. As he had been the model of a just and pacific king, so did he desire to be that of a victorious conqueror of nations, and, placing the regency in the hands of Isis, he went forth to war against Asia, accompanied by Thoth the ibis, and the jackal Anubis. He made little or no use of force and arms, but he attacked men by gentleness and persuasion, softened them with songs in which voices were accompanied by instruments, and taught them also the arts which he had made known to the Egyptians. No country escaped his beneficent action, and he did not return to the banks of the Nile until he had traversed and civilised the world from one horizon to the other. All this is legendary, but may contain a good deal of truth.

A year or two ago no one would have thought that the soil of ancient Egypt contained abundant relics of a Neolithic race. But the researches of the indefatigable Professor Flinders Petrie have shown that such was the case. Of this, his latest and perhaps most interesting discovery, we must give a brief account. In the first place it may not be amiss to remind the reader that General Pitt-Rivers (formerly Colonel Lane Fox) had already discovered near Thebes undoubted relics of the men of the Older Stone Age in the shape of true Palæolithic weapons, like those of the Thames or Somme gravels. But Professor Flinders Petrie found many such in the home of Palæolithic man, 1,400 feet above the Nile bed, on the plateau through which the river has since cut or carved out its valley.

But to return to Neolithic times, the Professor and his party came upon the site of the old town of Nubt, on the edge of the desert, where the god Set, brother of Osiris, was once worshipped. It is near Denderah, and the discovery throws light on a passage in the 15th Satire of Juvenal. Now, within a quarter of a mile of this town they came upon the site of another town, and the discoveries they made there were of a novel and startling character. Everything was different; no hieroglyphics, none of the usual Egyptian pottery, no beads or amulets, or scarabs, such as are found in every other town of Old Egypt. They also came upon a series of cemeteries, and nearly 2,000 graves were excavated. There were no mummies here; the bodies had been simply buried in the 'contracted position,' with the knees bent up to the arms. Evidently an inferior race had lived here; but who were they, and when did they flourish? This problem was settled by Mr. Quibell, one of the party, who showed that they were a Neolithic race who overthrew the high civilisation of the Old Kingdom, and ruled during the seventh and ninth dynasties, or about 5,000 years ago. Their advent caused a blank in Egyptian history which had been a puzzle to archæologists. This gap is now partly filled up. They were a fine powerful race, but very barbarous. They were cannibals, with a stature of six feet or more, and probably allied to the Amorites and Libyans.

Plate X.

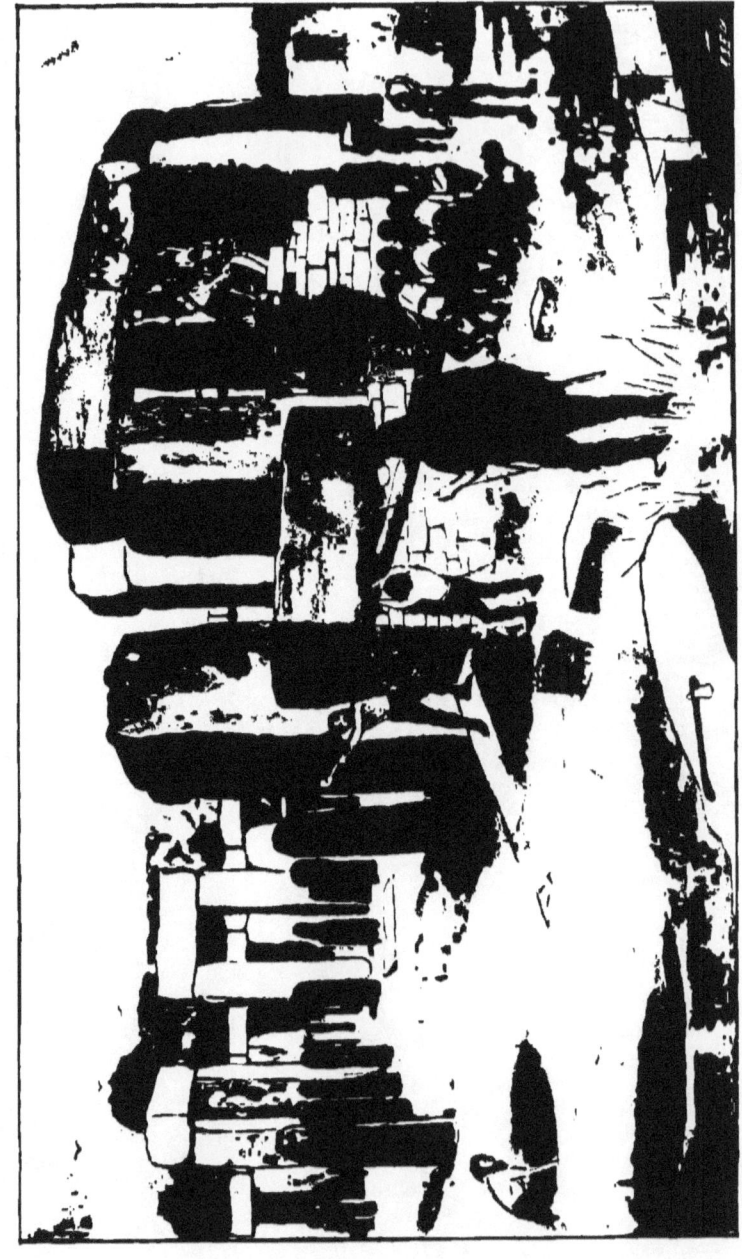

THE CONSTRUCTION OF STONEHENGE

## CHAPTER XI

### RUDE STONE MONUMENTS

*What mean ye by these stones?*—JOSHUA iv. 21.

THE traveller in the plains of Brittany and in the valleys of the Pyrenees encounters, at almost every step, strange monuments generally constructed of one or more unhewn stones of colossal size placed horizontally upon two, three, or four upright blocks, and sometimes on heaps of unmortared stones. Some of these primitive monuments were covered with earth, but probably not all of them. These are the dolmens, known also as stone chambers, Druidical altars, or by other fanciful names, and often incorrectly called cromlechs.[1]

Dolmens are undoubtedly tombs, and it is interesting to note that in many instances they are, or once were, surrounded by a circle of stones. Hence, possibly, the confusion that has arisen between cromlechs or stone circles and dolmens. The long chambered barrows spoken of in Chapter IX. were also encircled with standing stones, and in some cases chambered mounds and 'passage-graves' present a decided likeness to dolmens, only they have been covered up with earth or stones.

---

[1] The term 'cromlech' must not be applied to the above, for it is derived from 'Crom,' a circle, and 'Lech,' a stone, and so means a stone circle. Dolmen comes from 'Daul' or 'Tol,' a table, and 'man,' a stone, and so means 'table-stone.' The word man, maen or men, frequently occurs in Wales, Cornwall, and Brittany; *e.g.* Man-chester, Pen-maen-mawr. Tol is also a Cornish word. We get the word 'maen' again in the 'Menhir,' or standing (monumental) stone of the Khasia Hills in Northern India.

R

As Fergusson suggests, a chief may have had his dolmen constructed during his lifetime, to be covered up when the time came to bury him. Of stone circles proper we shall speak presently; but meanwhile it is interesting to note that many of the British stone circles are distinctly sepulchral, and the same is true of other countries. Thus, Palmer, in his 'Desert of the Exodus,' speaks of huge stone circles in the neighbourhood of Mount Sinai, some of them measuring 100 feet in diameter, having a cist [stone coffin] in the centre covered with a heap of large boulders. Kohen, a Jesuit missionary, saw in Arabia three large stone circles described as being very like Stonehenge, and consisting of lofty trilithons.

There are very few dolmens in Great Britain—unless we include megalithic chambered barrows under this head. A well-known example is Kit's Coty House, between Rochester and Maidstone, consisting of three large upright stones supporting a capstone of 11 feet by 8.

On the Continent the term dolmen is almost universally applied to the whole construction, including the covering mound or cairn. Thus French and other writers speak of a chambered mound or tumulus as a dolmen. But since it is probable that some never were covered up it seems better to make a distinction, as we do in this country. In France there are said to be about 4,000 dolmens, many of which would in England be called chambered tumuli. The Indian dolmens which are not covered up resemble those of Western Europe. Captain Meadows Taylor [1] examined a large number in India, and obtained particulars of no less than 2,129 in the Dekkan. About half of them had an opening on one side, probably for the free entrance or exit of the soul (as people thought then), just as in the Egyptian pyramid there was a passage to the chamber containing the mummy. With regard to the distribution of these structures, it is said that none are to be found in Eastern Europe beyond Saxony. They reappear in the Crimea

[1] *Trans. Roy. Irish Academy*, xxiv. 329.

and Circassia, whence they have been traced through Central Asia to India. They have also been noticed by travellers in Palestine, Arabia, Persia, Australia, the Penrhyn Islands, Madagascar, and Peru.

Their distribution in Europe is very irregular. From the Pyrenees they can be traced along the north coast of Spain, and through Portugal to Andalusia, where they occur in considerable numbers. Crossing into Africa, we find large groups of them in Morocco, Algeria, and Tunis. General Faidherbe examined five or six thousand at the cemeteries of Bou Merzoug, Wady Berda, Tebessa, Gustal, &c.

Years ago all such monuments were attributed to the Keltic people, because, as we shall show later on with regard to stone circles, archæologists were hardly aware of the former existence of pre-Keltic races. So everything that was very old and mysterious was put down to the Druids, and dolmens were often called 'Druids' altars,' in spite of the fact that they would have made very poor altars. Moreover, 'cup-markings' and other primitive engravings when found are almost always on the *inside* of the capstone or other stones, as, for example, at the dolmens of Keriaval, Kercado, Dol du Marchant, Gavr'innis (Morbihan), and the great tumulus at New Grange (Ireland).

But the Druidical theory, so popular since the time when Aubrey and Stukeley wrote, has been abandoned—though, unfortunately, it lingers on in popular guide-books. One fact alone is fatal to this theory—viz. that the geographical distribution of the 'rude stone monuments' (dolmens, stone circles, menhirs, alignments, or rows of standing stones) fails to correspond with the ethnographical distribution of the Kelts. In Europe, these occupy an elongated stretch of territory on its western sea-board, extending from Pomerania to North Africa. Now this area crosses at right angles the land supposed to have been occupied by the Keltic people on their westward course of migration.

It is difficult to believe, as some writers contend, that the dolmen builders were all of one race, nor can it be said that their migrations have been properly traced. According to Bonstetten, these people, starting from the coasts of Malabar, entered Europe through the passes of the Caucasus. Thence they spread themselves along the coasts of the Black Sea as far as the Crimea, where they divided, one stream directing its course towards Greece, Syria, and perhaps Italy and Corsica, the other northward, sweeping round the Hercynian forest. Later on, these wandering tribes penetrated into Brittany and Normandy, whence they overran the British Isles, advanced towards the south of Gaul, crossed the Pyrenees, and traversing Spain and Portugal obliquely, crossing the sea, they spread over the northern coast of Africa, and established themselves on the Egyptian frontier in ancient Cyrenia. But others have traced out quite a different course for them. Skeletons in the 'contracted position' are often found in the dolmens—a fact which is fatal to the Keltic theory, for, as we have seen already, this position of burial (without cremation) is highly characteristic of the Later Stone Age, or Neolithic state of culture. Once more let us remind the reader to be careful to remember that the terms Stone Age, Bronze Age, and Iron Age are not definite periods like the reigns of kings and queens, but represent states of culture. Doubtless those wise and learned Egyptians who raised the pyramids and built the older temples were using both bronze and iron, while other races in Europe had not progressed beyond the Neolithic culture-stage.

It may interest the reader to know that Worsaae has constructed the following chronological scheme for the North of Scandinavia :—(1) The Early Stone Age (Palæolithic), at *least* 3,000 B.C. (2) The Later Stone Age (Neolithic), about 2,000 to 1,000 B.C. (3) The Early Bronze Age, about 1,000 to 500 B.C., when the Stone Age was still going on in the North and an Iron Age (or culture-stage) had already come into the South. (4) The Late Bronze Age, about 500 B.C. to the time of the Birth of Christ when a pre-Roman Age of Iron was developed in Central and Western Europe. (5) The Early Iron Age, from 1-450 A.D., when bronze was still in use in parts of Scandinavia. (6) The Middle Age of Iron, about 450-700 A.D., when foreign Roman-German influence predominated. (7) The Later Iron Age, or Viking[1] period, about 700 to 1,000 A.D., when a Stone Age still lingered on in the extreme north of Finland and Lapland.

Cremation was not unknown to the people who constructed dolmens, for calcined bones are often found therein, and so are burial urns. The dolmen is obviously the prototype of the square box-like stone tomb of the last century. Much might be said about the weapons, ornaments, and utensils of the dolmens, but it will be sufficient for our purpose to record the following facts. They contain, besides the urns, drinking cups and other vessels of tolerably fine clay made by hand without the wheel, lance and arrow-heads of various kinds of stone, often beautifully wrought. Polished stone axes are rare in the centre and South of France, but fine specimens have come from Brittany, Denmark, and Sweden. Passing on to ornaments, we find that dolmens have yielded beads for necklaces of amber, jet, serpentine, turquoise, slate, alabaster, and several kinds of shells for ornaments. Nor are metals absent, for explorers have found bronze heads, pendants, bracelets, and axes. Pure copper also has been found. Iron only occurs in the dolmens of Algeria. With regard to the arts, these people evidently were not far advanced, they could not write—not even picture-writing being known to them, much less the alphabet. They have left only a few simple engravings on their tombs, such as a pair of broad feet in outline—suggestive of some prehistoric 'Trilby!' Sometimes they engraved pictures of axes, and at other times only vague patterns, such as concentric half circles (Gavr'innis) like big necklaces, which is probably what they were meant for.

[1] The word 'Viking' has nothing to do with kings, and should be pronounced Veek-ing, from Vik, in the south of Sweden. Probably Wick, in Caithness, is the same name, as there are many Scandinavian place-names all round the Scotch coast.

Most writers attribute the dolmens to an Iberian race which preceded the Kelts and lived on after the Keltic and the Roman conquests. But in the more northern parts of Europe some may be attributed to the Finns, spoken of in the previous chapter.

The various kinds of 'rude stone monuments,' such as dolmens, standing stones, or menhirs, avenues and alignments of such, and stone circles, are found throughout a large part of the Old World, and some even in the New. According to General Pitt-Rivers, they are continuous, or nearly so, from the Khasia hills of Northern India to Central Asia, Persia, Asia Minor, the Crimea, along the north coast of Africa, bordering the Mediterranean, in Etruria, up the south and west coast of France, into Britain, and as far as Denmark and Sweden. As far as present knowledge goes, they are unknown in Russia Proper, in Northern Asia, Central and South Africa, and on the two American continents, with the exception of Peru.

We pass on now to a brief account of each of the above groups of stone monuments, and standing stones, either singly or in groups, must claim our attention first, for they are the simplest and probably the oldest form of memorial or monument. But they were used for various purposes, as a reference to written history amply shows. To take the sacred record first, we find in Scripture more than one allusion to standing stones, and there we learn some at least of the purposes for which they were erected. When Rachel died, Jacob 'set a pillar upon her grave' (Gen. xxxv. 20), and in the time of Samuel her sepulchre is referred to as a well-known place (1 Sam. x. 2). When Jacob and Laban made a covenant between themselves, the former 'took a stone and set it up for a pillar,' and, surrounding it with a cairn of stones, called the place Galeed, or the heap of witness (Gen. xxxi. 47); and we read in verse 51 that 'Laban said to Jacob, Behold this heap and behold this pillar which I have cast between me and thee, that I will not pass over this heap to thee, and that thou shalt not pass over this pillar to me to do me harm.' Jacob also erected a stone at Bethel to record a certain important event (Gen. xxviii. 18). At Mount Sinai, Moses set up twelve pillars (Exodus xxiv. 4). A stone had been erected over Bohan, the son of Reuben, which afterwards appears to have been recognised as a boundary (Joshua xv. 6 ; xviii. 17). Joshua erected a pillar under an oak for a religious purpose, and as a witness against the people (Joshua xxiv. 26, 27). Again, when the Israelites had crossed over Jordan, Joshua took twelve stones and pitched them in Gilgal. 'And he spake unto the children of Israel, saying, When your children shall ask their fathers in time to come, saying, What mean these stones? then ye shall let your children know, saying, Israel came over this Jordan on dry land' (Joshua iv. 21, 22). Absalom was buried under a heap of stones, and we are told that 'he reared for himself a pillar which is in the king's dale ; for he said, I have no son to keep my name in remembrance, and he called the pillar after his own name, and it is called unto this day Absalom's Place.'

We learn from early Irish manuscripts descriptive of ancient cemeteries and battle-fields in various parts of Ireland that the memorial of the Keltic warrior and chief was a cairn and pillar stone. Cormac Cas, ancestor of the O'Briens, was buried under three pillar stones, which gave the name of Dun-tri-liag to a fort which he erected. But these customs are evidently a survival from older times, in fact, from the Neolithic Age.

From various records it is clear that long after the first introduction of Christianity into Europe, the pagan population, and those who were only partially Christianised, clung with great pertinacity to the worship and veneration of rude stone monuments. The decrees of the Councils show that in France they were objects of veneration down to the time of Charlemagne.

A decree of a Council at Nantes exhorts 'Bishops and their servants to dig up, remove, and hide in places where they cannot be found, those stones which in remote and woody places are still worshipped, and where vows are still made.' In Scotland there yet remain a great many single standing stones; it has been said that examples might be quoted from almost every parish in the country! They are so ancient as to have outlived even the traditions of those who set them up, so that later traditions have often been invented to account for them—a fact which tells strongly against their having ever been connected with the Druids or Keltic priests. The Hare Stane (stone) on the Borough Moor of Edinburgh, celebrated in the lay of Marmion, still remains. Probably it marks the western boundary of the ancient chase claimed from time immemorial by the neighbouring capital. What the word 'Hare' or 'Hoar' means is not known, but in other cases such stones appear to be sacred, and in memory of the departed. The Camus Stones probably served as landmarks. Cat Stones (from Cat=battle) appear to commemorate some fight. There are also Kings' Stanes (cf. Kingston-on-Thames), Witches Stanes, and Tanist Stanes. The Witch Stane near Cairnbeddie, in Perthshire, is associated with local traditions rendered classic by Shakespeare, for here it was that Macbeth is supposed to have met by night two celebrated witches. The Hawk Stone, or *Saxum Falconis*, at St. Madoes, Perthshire, still bounds the parishes of St. Madoes and Inchture, and is said by Boece to have been set up after the defeat of the Danes at the battle of Lunearthy, about A.D. 990, but it is much older than that. Then there is the Tanist Stone,[1] used for ceremonial purposes on the accession of kings and chiefs. The most celebrated of these is the Lia-Fail, formerly at Scone, and now at Westminster Abbey, where it is part of the coronation throne. This venerable relic of a remote age is believed to have served for many ages as the coronation throne of the monarchs of Ireland. It was removed to Scotland and deposited at Iona, or Icolmkil, for the coronation of Fergus Mor Mac Eare, a prince of the blood-royal of Ireland. Then it was translated from Iona to the Abbey of Scone, when the Scottish kings had extended their sovereignty

---

[1] Gaelic, *Tanaiste*, a thane, or lord; the next heir to an estate.

over the ancient kingdom of the Picts. In Saxon Scotland it bore the name of 'The King's Stone,' and was regarded as the national palladium, until in 1296 Edward I. had it brought to Westminster Abbey, as an evidence of his absolute conquest of the kingdom.

From the earliest times standing stones appear to have been used as the most sacred witnesses of every solemn covenant, including that between the elected chief or king and his people. The use of stones in this manner can be traced to the East and goes far into the remote past. Holy Scripture confirms this, for when Abimelech was made king, it was by the pillar which was in Shechem (Judges ix. 6); and when Jehoash was anointed king by Jehoiada, the king 'stood by a pillar as the manner was.' Some old deeds bear witness to the veneration in which stones and stone circles were held. Thus, as Dr. George Petrie has pointed out, in the year 1349 William de Saint Michael was summoned to attend a court held 'apud stantes lapides de Rane en le Garniach [Orkney],' to answer for his forcible detention of certain ecclesiastical property; and in 1380 Alexander, Lord of Regality of Badenoch, and son of Robert II., held a court 'apud le standand stanys (stanes) de la Rathe de Kyngucy Estir,' to inquire into the titles by which the Bishop of Moray held certain of his lands. As the present writer well knows, there are no standing stones remaining at this beautiful and popular Highland resort (now spelt Kingussie; it lies in the broad Spey valley, and was greatly loved by the late Professor Blackie). A rathe or rath was a fortified mound or hill, and possibly the one here mentioned is the hillock on which the Established Church is built.[1]

In France over 1,600 examples of menhirs have been recorded,

[1] On writing to our friend Mr. A. Macpherson, of Kingussie, author of a very interesting work on the later history of the district, entitled *Glimpses of Church and Social Life in the Highlands in Olden Times* (*q.v.* pp. 121, 122), we received the following reply: 'The word "Estir" means simply Easter. So far as I have been able to trace, the standing stones stood on the little hillock on which the present parish church of Kingussie is built. The spot was known in olden times as *Tom a Mod*, or the place of meeting or gathering.'

of which about half are in Brittany. At Locmariaquer (Morbihan) is one which is said to be the largest in the world. It is in the form of a rude obelisk of granite, brought from a distance, and lies on the ground broken into four pieces, the whole length of which is about 67 feet, and weighs about 342 tons. One at Plésidy (Côtes-du-Nord) measures about 37 feet in height. Then there are the world-famous alignments at Karnac, of which we shall speak presently. In England monoliths are often associated with stone circles—*e.g.* the King's Stone at Stanton Drew, Long Meg at Little Salkeld, the Ring Stone at Avebury, and others. One of the finest monoliths stands in the churchyard of Rudston, Yorkshire. Large examples are met with in Algeria, Morocco, India (Khasia Hills), and Central Asia.

There is reason to believe that the Keltic people, on becoming Christianised, sometimes converted the menhirs or standing stones of a previous race into crosses, and this may account for the very short arms of the beautiful Keltic crosses. All lovers of art must be glad to see that copies of these ancient monuments are to be found in many of our modern churchyards, with their lovely waving and interlacing patterns; but we notice with regret that some people prefer to have modern and far less beautiful designs carved thereon, whereby their chief charm is lost.

With regard to alignments, the finest examples of such are to be found in Brittany, in the vicinity of Carnac. They occur in groups within a few miles of each other, at Ménec, Kermario, Kerlescant, Erdeven, and St. Barbe. Of these the first three are considered to be portions of one original and continuous series, extending for a length of nearly two miles in a uniform direction. The menhirs, commencing at the village of Ménec, are arranged in eleven rows. Of these, some are from 10 to 13 feet high, others only three to four feet. Then follows a group at Kermario with ten rows. After another interval they appear again at the village of Kerlescant in thirteen rows. Concerning those at Erdeven, it has been shown that out of a total of 1,120 menhirs, only 290 are still standing, 740 have fallen, and 90 have been removed.

Here they can be traced for nearly a mile, but the stones are smaller than at Carnac. Altogether, about fifty alignments are known in France. It is important to notice that alignments are often connected with other monuments, such as stone circles, tumuli, &c. Thus, at Carnac, the first three of the above groups begin with and strike away from some kind of monument. At the head of the Ménec division is an enclosure of small stones set close together; when complete it apparently joined the centre row of stones in the alignment. At Kermario a dolmen stands in front of the alignment in a conspicuous position. At Kerlescant is a quadrangular enclosure mostly composed of small stones set closely together, while on the fourth side is a long barrow. Near the Erdeven alignment is also a tumulus.

At a place called Penmarch there is an alignment containing 200 menhirs arranged in four rows. Others, with only a single row, are found at Kerdouadec, Leuré, and Camaret. Alignments are found in other countries. In the Pyrenees they are generally in single file, and mostly straight. In Britain, the only example of a single row is that known as 'The Nine Maidens,' at St. Columb, in Cornwall. In our country they generally take the form of avenues leading to or from other monuments, each avenue consisting of two rows of stones, sometimes called 'parallelitha,' as at Avebury, Stonehenge, Shap, Callernish, and on Dartmoor. In the Vale of the White Horse are some alignments reminding one of those at Carnac, consisting of 800 stones grouped in three divisions over an irregular parallelogram.

Before passing on to consider stone circles let us endeavour to answer the question put at the head of this chapter—'What mean ye by these stones?' No one, in the present state of archæology, can pretend to give a full, final, and complete answer to this question; but we can at least put before the reader certain facts, customs, and traditions which will be of great assistance in helping him to form an opinion ; and at the same time we will take the opportunity of considering the methods by which big blocks of stone may be transported and set up. This subject, so frequently spoken of as a great mystery, is, after all, a comparatively simple one, and the difficulties with which primitive men had to contend in conducting such operations have been much exaggerated. For this purpose we cannot do better than put before the reader in brief form some highly interesting facts collected by Major Godwin-Austen [1] in the Khasia Hills of Northern India, where a very primitive people still linger on from remote antiquity. 'Certainly the most striking objects of interest in the Khasi Hills are the upright stone monuments that are to be seen all over the country ; these, set up by the wayside or in the villages, more frequently cutting the sky on prominent hills, with the large slabs horizontally set before them, at once recall the Druidical remains [so called] of our own island, Northern France, &c., and lead one to marvel at the similarity of the custom, and to inquire into its origin and design. Many who visit those hills take it at once for granted that they are the graves of illustrious men ; or, after a vain endeavour to get some information from the coolies about them, let the matter rest ; or finally believe that the ashes of the dead, to whose memory the monoliths are erected, are buried under the flat kind of altar or dolmen seen in front.' Major Godwin-Austen then reminds the reader that it is very difficult to obtain information from a semi-civilised people about their religious ideas, especially when one speaks through an interpreter. The tall upright stones are called *Mao bynna*, from *Mao*, stone, and *bynna*, to make known, to be informed, literally a monument. There is a flat stone in front of the upright ones, and the ashes of the dead are *never* deposited there. As might have been expected, these monuments have no connection with funeral rights, but are erected solely in order to perpetuate the memory of a person long

[1] *Anthropological Journal*, vol. i. p. 122.

deceased, who, as a spirit, has brought good fortune to a descendant, his family, or his clan.

Wealth or renown have no connection with the size of the stone, but that depends on the wealth of the person who erects it, and on the supposed benefits conferred by the deceased after passing into the world of spirits and demons. According to the belief of these primitive hill people, the spirits of the dead and demons are the cause of all joy or woe ; these can give riches, or inflict death and disease.

Major Godwin-Austen says : ' The history connected with the erection of some large slabs near Cherra Poonjee will exemplify this curious custom. One of the clans " Kūr," in Cherra, is known as the "Nongtariang," and many years ago died an old lady of the clan, not famous for anything in particular during her lifetime, but whose virtues appear to have been great after her demise, for after this the Nongtariang clan, from being a poor one, rose gradually to considerable wealth. She, when propitiated and called on for aid, never failed her race, and in return, after some sixty years or more, they raised to her memory five well-cut stones, which are to be seen on the west side of the road between Cherra Poonjee and Suraran, the central monolith adorned with a kind of rose cut in relief on the front face, and an ornamental disc on the apex. It would appear that she still remained the guardian spirit of her clan ; they continued to prosper, and as a further token to her memory they added in 1869 five more stones on the other side of the road, and in a line with the first set.'

But how are people to discover what their ancestors are doing, or have done, on their behalf? To the Khasi people this hitherto insoluble problem appears quite simple, they seek omens by the breaking of eggs ! If the benefit conferred is not great, only stones of small size are set up, but if, on the contrary, it is some great favour, then large monoliths are put up, such as form conspicuous landmarks on many a hill-top.

But in some cases the monument has been set up as a thank-offering for recovery from illness. 'During the illness of a person every kind of propitiation and exorcism, either by the breaking of eggs, or sacrifice of fowls, pigs, &c., and by the examination of the liver and viscera, having been made and then failing to restore him, the sick man may vow that should he recover he will erect a set of stones to one of his ancestors, who, it is presumed, on knowing of the intention, will do his best to save him.'

In the erection of these monuments certain socialistic principles are brought into action, for all the members of the community are obliged to assist—and no doubt are quite willing to do so. Here is co-operation of a very practical and brotherly kind! All they receive is a little food or drink at the family dwelling in the evening. But an exception is made in favour of the skilled workmen who cut the stones. The family also provide music, and the beating of tom-toms is kept up all the while. Most of these monuments are of great age, for their history is lost.[1] The finest examples occur in the central portion of the Khasia Hills and near the larger villages. The largest collection of huge slabs and upright stones seen by Major Godwin-Austen was at Lailang-kote; but these, he considers, were erected for a very different purpose, and suggests that it was a meeting-place for chiefs and elders of clans. There are steps leading up to the huge platform, which consists of one huge stone slab weighing over twenty tons. He was not fortunate enough to see people in the act of raising a stone, but he saw the spars which had formed a sort of cradle on which big stones had been dragged. They were strong curved limbs of trees roughly smoothed and rounded.

[1] Among other writers on this subject are—Mr. John Eliot, in *Asiatic Researches*, vol. iii.; the Rev. A. B. Lish, in the *Calcutta Christian Observer* for 1838; Dr. Hooker, in his *Himalayan Journals*, which contains good drawings of the monuments; Rev. W. Pryse, in the above newspaper, March 1852; Dr. Thomas Oldham, of the Indian Geological Survey, on *The Geology of the Khasi Hills, with Observations on the Meteorology and Ethnology of that District*.

With regard to the religion of these people, he says :—' In their funeral rites there is much that is wonderfully strange, and, to those unacquainted with their superstitions, appears unmeaning. There can be no doubt that the Khasis have a very strong belief in a future state, and to say that they have no religion, as has been stated by some writers, is erroneous. True, they have no temples for either worship or idols; the religion of the people is principally a demon worship, and we find the Khasi ever in communion with, or in the power, as he supposes himself, of the spirits of those who have gone before him; while, added to these spirits of their ancestors are numberless demons, male and female, ever ready, if not propitiated, to bring evil upon himself or his undertakings, and whose power even extends over the spirits of the dead. Every dark shady wood, every stream, every conspicuous hill has its presiding demon, which, in many instances, gives the name to the site. Thus, the hill of Lārū, in North Jantia, beyond Nongjinghi, is the abode of a one-legged demon, whom to see is death, but who was occasionally heard in the dense forest that clothed the northern steep face of the mountain, so said the credulous villagers of Nongtūng.'

The monuments we have been considering have nothing to do with funeral rites. With regard to burials, they have a kind of dolmen, where the ashes or bones of the dead are deposited after cremation. These are collected and placed in an earthen vessel, and buried somewhere near the place where the burning took place, a stone being placed over the vase for security. Then, after about a year or more, the remains are placed in the dolmen, which belongs to a clan or family. The survivors must first ascertain whether the soul of the dead man or woman is at rest, and leading a happy existence. This is supposed to be the case if all the family are in health. In these vaults or dolmens the bones are collected with the idea that the souls of the departed may all mingle together. But man and wife never rest together, because they belong to different clans.

The idea of a member of a family being a wanderer in the other world, cut off from and unable to join the circle of the spirits of his own clan, is most repugnant to the feelings of a Khasi or Sinting. This is the reason why, on the death of a man in a distant village or district, every attempt is made to lead his spirit back, with his calcined bones, to his native place. When carrying the bones home (see Chapter IX., p. 211), great care is taken of them, and they do all they can to prevent the man's spirit wandering away. Like the Greeks, they believed the dead to be but feeble ghosts; they could not, therefore, have the strength to cross a river; and so the people carrying the bones stretch a thread of cotton right across it. This they call 'The String Bridge,' and the idea is that the spirit or ghost can glide along it! All this may seem to modern Europeans very childish, but, at least, it may be said that such people think more about, and commune oftener with, their departed ones than we do.

> How pure at heart and sound in head,
> With what divine affections bold,
> Should be the man whose thought would hold
> An hour's communion with the dead!
>
> *In Memoriam.*

The Rev. J. Pryse, in his excellent paper,[1] referred to above, says: 'The universal desire to immortalise the memory of their dead ancestors on earth by the erection of stone monuments may be deemed as a faint indication of an expectation of some kind of future existence. The sacrifice for their dead, which they called 'Suid-căp,' that is, the ghost or demon of the dead, may be considered indicative of the same notion. The professed intention of the sacrifice is to pacify the spirit of the dead, so that it may not come in the capacity of a 'ksuid,' or demon, to cause pain and calamity to the family. This sacrifice is frequently repeated after a person's death, if his bones were deposited

---

[1] *Calcutta Review,* 1854, p. 128.

in a small repository; but if they are placed in a large one, the fear of his injuring the family is not so great, and the sacrifice is not so frequent, because ... the religion and customs were observed regarding him.[1]

We give below the form of words used at the sacrifice.[2]

These hill tribes have a remarkably simple and effective way of transporting big stones, as the following account, published by Mr. A. L. Lewis, will show :—[3]

'A stone having been selected from some place where there are natural cracks, into which levers and wedges may be introduced, is split from the parent rock by those instruments, and moved on rollers till its weight is transferred to two or three

---

[1] See also a paper by Mr. S. B. Clarke, *Anthrop. Journal,* vol. iii. p. 481.

[2] '*Thank Offering to Secure the Protection of the Goddess.*

'Attend, O Goddess, that thou mayest protect us all as a family, that we may increase and prosper. O Goddess, we are now about to sacrifice a large cock, and a large-horned goat, also a plantain leaf, flour, baked rice, a heart, and a twig. Do thou, O Goddess, attend to us. I put up and cause to stand the offering for the purpose of appeasing thee, O Goddess, thou family-builder. I remove and drive away thirty demons, nine demons [*i.e.* an indefinite number], and Ka Tyrût Ka Smer [a powerful and malignant demon], that I may prepare and set aright the intestines, that I may cut the throat of the cock, and sprinkle the blood upon the twig. Away, thou Ka Tyrût Ka Smer! I sweep away and remove thee, that I may open a clear road for the purpose of inquiring concerning the good and the evil. Thus it is proper that I should sacrifice a large-horned goat. Attend thou, O Goddess, whilst I offer to thee a member and the backbone, after I shall have laid it bare and made it ready. Attend thou, whilst I observe and keep the rules and customs, and whilst I offer to thee a sacrifice in order that thou mayest give us health, that we may increase and prosper, that we may walk securely, that we may enjoy our possessions in security, that we may carefully ameliorate our families, that we may increase in number. Do thou embrace us, do thou confide in us, do thou support us while we observe the rules and customs, that we may become numerous, that we may offer the sacrifices. Come thou and receive us, receive us at thy feet, that we may spread out on the right and on the left. Come, confide in us; come, support us; come, envelop our spirits with thy power.' The sacrifice is generally made by a whole kūr, or family, all related by blood.

[3] *Anthropological Journal,* vol. viii. p. 182. Mr. Lewis gives it in a paper of his on certain stones near York, called 'Devil's Arrows,' where he states that the account was given by Mr. Greey, C.E. (since deceased), to the late Dr. Inman, who sent it to him for publication.

straight tree trunks cut for the purpose; under which strong bamboos are placed crosswise, which again rest on a number of smaller bamboos, and these again upon others; if the stone be very large, the smallest being far enough apart to allow a man to stand between them. All these being lashed together at each crossing, form a simple but substantial framework, which may be made of such size as to allow a sufficient number of men to grasp, lift, and transport it and its burden, so that a stone weighing 20 tons has been known to be carried up a hill 4,000 feet high in a very few hours.

'It has been calculated that three or four thousand men could in this way transport either of "The Devil's Arrows" any distance that might be wished. On reaching the spot where the stone is to be erected a hole is dug of sufficient depth to keep it steady, into which one end of the stone is allowed to slide. Ropes are then attached to the framework, on which the other end still rests, and by hauling at them the stone is quickly set up.' We shall speak of the way in which big stones may be raised in our next chapter on Stonehenge.

Stone circles, known as dom-rings, or thing-steads, in Scandinavia, law-tings in Orkney, and cromlechs in France, are usually from 20 to 100 feet in diameter. Among Northern nations they were used in later times for the assembling of the people at the election of their princes. For example, Eric, King of Sweden, was elected at a celebrated monument of this kind at Upsala, in the year 1396. There is no doubt that among the Gothic people they were used as courts of justice. Great veneration was attached to them, doubtless from their high antiquity, and from the fact that stones were formerly objects of worship. Many churches have been built on the sites of stone circles. There is a common Gaelic phrase, '*Am bheil thu dol don clachan*'—'Are you going to the stones?' that is to say, 'Are you going to church?' They were also battle-rings, wherein trial by single combat took place.¹ The Gaelic word *clachan* signifies both a circle of stones and a place of worship.² Even so late as the year 1438 we find a notice that 'John off Erwyne and Will Bernardson swor on the Hirdmane Stein before oure Lorde ye Erle off Orkney and the gentiless off the cuntre.' It is, perhaps, not inappropriate here to remark that in both the 'Iliad' (xviii.) and the 'Odyssey' (viii.) of Homer assemblies of elders are mentioned as sitting in solemn conclave on stone seats arranged in circles. Such a fact as this naturally suggests the question whether the chapter-houses of our cathedrals, which are round, may not represent a survival of this ancient custom.

Sometimes a stone circle consists of mere boulders rolled into place; in other cases they are formed by stone pillars, chosen for their length, and inserted in holes in the ground. They are not

---

[1] See *The Saga of Skallagrimson*, chap. lxxxvi., by Rev. W. C. Green, where such a trial is described.

[2] The largest stone circle in France stands on the Ile-aux-Moines (Morbihan), in the village of Kergonan, and when complete had a diameter of 100 metres (328 feet); but a few of the British examples are larger, among which may be mentioned Avebury, 1,260 feet by 1,170 feet; Stonehenge earth circle, 300; Stanton Drew, 360; Brogar (Orkney), 345; 'Long Meg and her Daughters' (Westmoreland), 330. One near Dumfries, called 'The Twelve Apostles,' is nearly 100 metres. One at Mayborough, near Penrith, is nearly 300 feet in diameter. Stone circles were certainly sacred. We shall speak of their probable use presently. With regard to the worship of stones in India, the practice has not even yet entirely died out, but it exists chiefly among older and non-Hindu races. Groups of standing stones may be seen in that country, of which each one stands for some god. In Southern India five stones are often to be seen in the ryot's field, placed in a row and daubed with red paint; these they consider the guardians of the field. Even in the fourth century B.C. Theophrastus refers to the custom of anointing stones. Compare the following passage from Isaiah:

> Among the smooth stones of the valley is thy portion;
> They, they are thy lot:
> Even to them hast thou poured a drink-offering,
> Hast thou offered a meat offering.

always single circles, for sometimes we find several concentric circles within the outermost one. There is generally a trench, or trench and rampart, outside, and a pathway, or avenue, leading to the interior. The numerous small circles of small stones close together are probably burial-places. These are sometimes nearly covered by turf or peat. Some of the Scotch circles have been excavated, and in about twenty cases burials of the so-called Bronze Age, mostly by cremation, have been discovered within them, and cinerary urns, with the characteristic ornamentation of that period or culture-stage, also were found. The stones appear always to be placed at equal distances, and there *may* be some significance attaching to the number thereof. To give a few examples: According to Dr. Thurnam, the two inner circles at Avebury, the lesser circle at Stennis, and one at Stanton Drew (near Bristol), each consisted originally of twelve stones; the outer circles at Avebury, the outer one of trilithons at Stonehenge, the large circle at Stanton Drew, and the circle at Arbor Low, each consisted of thirty; those of Rollrich and Stennis of sixty stones; and the large enclosing circle of Avebury, of one hundred stones. At Boscawen and adjacent places in Cornwall, four circles have each been formed of nineteen stones. But no one at present has been able to show what these numbers meant, so perhaps it will be as well not to expect any important results in that direction. Let us first consider some of the Scotch stone circles, and see what evidence they afford of the original purpose, or purposes, of these monuments.[1]

In a remarkable circle at Crichie, in Aberdeenshire, a central cist (or stone coffin) was found surmounted by a pillar, and near the base of every pillar in the circle was found some deposit, such as an urn, or bones, or some axes. At another circle at Tuach, in the same neighbourhood, urns, with deposits of burnt bones, were found around the base of a central pillar, as also near most of the other pillars forming the circle.

[1] See *The Sculptured Stones of Scotland*, by John Stuart, pp. xxii to xxx.

Among other examples mentioned by Mr. John Stuart in his big book is one at Moyness, in Nairnshire. There are three concentric circles here, the innermost being paved with small stones, and in the centre of the latter a clay urn of rude make was found. A large double circle at Wardend, in Aberdeenshire, when excavated, revealed an urn in the centre of the inner circle, and a deep layer of bones and burnt matter. At Callernish, in the Lewis (to be described further on), two rude stone chambers containing burnt bones were found near the base of a great pillar in the centre. It is interesting to note, as giving a faint clue to the antiquity of the monument, that peat had accumulated over the pavement to a depth of five to six feet.

On the west coast of the isle of Arran a group of stone circles on Manchrie Moor, Tormore, was explored, with the result that in the centre of each circle, and at other points of it, were found both short cists and rude urns. Another group of four circles in the parish of Banchory-Devenick, Kincardineshire, revealed burnt bones and portions of urns. Old accounts are often of great value. Thus, the great circle at Little Salkeld (Westmoreland), known as 'Long Meg and her Daughters,' was visited by Camden in the year 1599, and at that time there were two cairns within it, which have long since been removed (at least before Stukeley's time). Again, the entrenched circle at Mayborough, near Penrith, is said by Stukeley (1776) to have had two stone circles in its area, whereas now there is only one solitary pillar. One of the Stennis circles even now contains a ruined dolmen.

Turning to Ireland, Dr. G. Petrie[1] described a remarkable collection of stone circles, cairns, &c., at Carrowmore, about two miles from the town of Sligo, all of which have either kistvaens, or dolmens, within them. Some have a simple circle of stones, others two, or even three. They are all clustered together irregularly round a great cairn. Many are dilapidated, but

[1] *Proceedings Royal Irish Academy*, 1837-8, vol. i., p. 140. See *The Sculptured Stones of Scotland*, by John Stuart, vol. ii., p. xxiii.

remains of about sixty are left. Many were destroyed by the peasants, but in all those so broken up were found human remains, urns, &c., and one of them was filled with the bones of men and animals. The monument at Shap, in Westmoreland, when complete, consisted of a long avenue of pillars, apparently connected with circles and barrows. In Scandinavia such monuments take many curious shapes—circles, ellipses, squares, and are said to be burial-places of the Iron Age and probably of the Vikings.

The finest example in Britain is, or rather was, that of Avebury, or Abury, a small village in Wiltshire, a few miles from Marlborough; but very little of it is now visible, as country people have not much respect for the monuments of antiquity, most of the monoliths having been used for building! This wanton destruction, however, will not be possible any longer, thanks to Sir John Lubbock and the Act for the Preservation of Ancient Monuments. The outer circle once consisted of 100 large blocks, and has a diameter of about 1,200 feet (various dimensions are given by different writers). These stones had a height of 15 to 17 feet. This circle was surrounded by a broad ditch and lofty rampart; within its area were two smaller circles, with diameters of 350 and 325 feet. They were not placed in the axis of the big one, but a little to the north-east. Each of these consisted of two concentric rows of stones, a menhir occupying the centre of one, and a dolmen that of the other. The embankment, which is broken down in several places, had originally an entrance to the circle, from which issued a long avenue of approach known as the Kennet Avenue, consisting of a double row of stones, and running for a distance of 1,430 yards in a straight line. The oldest account we have is that of Aubrey, 1663, by command of King Charles II. Dr. Stukeley's descriptions and very fanciful theories about Stonehenge and Avebury were published in 1740. When Sir Richard Hoare examined the former in 1812, many of the stones had been removed.

It is interesting to note that in the tenth century Avebury was considered to be a place of burial (as its name alone would imply). An old charter of King Athelstan, dated 939, describing the boundaries of the Manor of Overton, in which Avebury is situated, says: 'Then to Collas barrow, as far as the broad road to Hackpen; thence northward up along the Stone row; thence to the burying-places.'[1] At Hackpen was another monument. Hackpen, or Hacaspen, was a double oval or circle of 138 feet by 155 feet, with an avenue 45 feet wide, which is supposed to have extended in a straight line for a quarter of a mile or more, and pointed towards Silbury Hill, the third member of the Avebury group. This is a great conical mound about a mile off, covering an area of five acres, and with a height of 130 feet! Its nature and age have been much discussed. It might be a great barrow, or a big hill formed for meetings, where a chief or king could meet his people, and hold courts of justice.

We must now say a few words about the circle at Callernish, in the Lewis, one of the most remarkable of all the British circles. Here we have a cruciform group of standing stones attached to a central circle with a diameter of about 40 feet. It is in a conspicuous place, and in the centre is a column nearly 17 feet high, round which is a circle of flat upright stones. From this an avenue of similar stones stretches nearly 270 feet to the north, while single rows placed towards the other cardinal points complete the arrangements. The plan of the whole group bears a striking resemblance to a Keltic cross. It probably had a length of 680 feet, but much of it is lost.

Through the liberality of Sir James Matheson, this and some neighbouring circles were explored, and at Callernish a rough causeway was discovered below the peat which had accumulated in the course of ages. The explorers also found a circular stone building on the east side of the central stone, which contained two chambers, the largest 6 feet 9 inches by 4 feet 3 inches, and here were human bones apparently subjected to the action of

[1] See Fergusson's *Rude Stone Monuments*, p. 73.

fire. This is a very important discovery, and one which connects the stone circles with chambered cairns. Probably this cairn was somewhat later than the circle in which it stands. Had the circle been of later date the builders might have made this monument their centre, instead of the pillar stone which they put in the centre.

Next in importance to the great stone circle known as Stonehenge, to be described in our next chapter, come the 'Stones of Stennis,' situated in the Isles of Orkney. In their rude unhewn simplicity they remind one more of the megalithic groups of Avebury, Stanton Drew, and other places, rather than of Stonehenge, with its symmetrical and highly-finished trilithons, the lintels of which were fixed to the upright columns by means of the mortice and tennon. No such skill is manifested here in this barren Orcadian promontory. No tools were used but such as were required to dislodge each of the great blocks from the quarry, for their shapes are only due to the natural lines of division in the rock itself. For this purpose implements of stone were amply sufficient; and although many writers have persuaded themselves that metal tools must have been required in the construction of all these wonderful monuments, yet we venture to assert that no impartial reader can possibly regard the question, which is one of vital importance, as settled in their favour. In fact, all the evidence at present collected seems to point quite the other way. Just to take one example, cases are on record of the discovery of polished stone axes in cavities in granite blocks which appear to be the actual implements used by the men of the Later Stone Age in breaking up such blocks, or in dislodging them from the quarry.[1]

The Ring of Brogar, as this circle is commonly called, appears to have originally been composed of sixty stones, of which, unfortunately, only twenty-three now remain, together with a few broken stumps. Except at two points, it is surrounded by a trench, with

[1] *Archæologia*, vol. vii., p. 414.

a diameter of 366 feet, and an area of nearly two and a half acres. A little way within the trench are the erect stones, the highest of which stand about 14 feet above the ground, while others show an average height of 8 to 10 feet. They are all from that now classic formation, the Old Red Sandstone. A smaller group of stones, known as Stennis Circle, is enclosed by a mound of earth instead of a trench, and once contained a large cromlech, of which the ruins still remain. The diameter of the ring originally formed by these stones is about 52 feet, and, judging from the space between those still standing, it would appear that they were twelve in number. It is possible that an avenue of standing stones originally led up to the eastern entrance to the larger circle, now indicated by a break in the trench. Great havoc has been made here, as elsewhere, of these precious memorials of a bygone age; for, besides the two circles above alluded to, there are to be seen prostrate blocks representing two others.

It has too long been the habit of English archæologists, when hard pressed for a theory to account for stone circles, not only to make a most unwarranted assumption and call them 'temples,' which was bad enough, but to bring in their old familiar friend the Druid as the builder of these supposed temples! Poor Druid, how amused he would be could he but read some of the hundreds of learned books, pamphlets, journals of societies, guide-books, magazines, daily papers, &c., in which he is invariably trotted out, like a *deus ex machina*, in order to come to the rescue of archæologists whose minds seem unable to wander further back through 'the corridors of time' than the days of our Keltic ancestors! Anyone taking up the literature of stone circles, dolmens, cromlechs, &c., will find the Druid cropping up everywhere, until—if he thinks for himself—his patience at last becomes fairly exhausted. But it is amazing to see how readily the Druidical theory is accepted by most people; nor is the reason for this far to seek. Unfortunately, our historians, and others who write text-books for schools, or historical tales for young people, seldom tread in the paths of unwritten history. Archæology to them is an unknown and unconquered country, and, however delightful to some of us these 'fresh woods and pastures new' may be, our historical friends, following their leaders like faithful sheep, refuse to feed therein, not from fear perhaps, but rather from ignorance, or, at least, mistrust. Had they but taken the trouble to inquire of some of our most distinguished archæologists, they would have experienced the pleasure of finding their mental horizon considerably enlarged, in learning for the first time of the existence of one or two races that lived in Britain long before the Keltic invasion ! And so it is that one's earliest recollections of history lessons, either at home or at school, are associated with the mistletoe, wickerwork cages for human victims, long white robes, and other paraphernalia of the Druid. And naturally one is brought up with the idea that Druidical times are the *Ultima Thule*, or extreme limit of history; but it is a thousand pities that, in the face of so much striking evidence, both from tradition and from archæology, as well as written history, schoolmasters and others should persist in putting the Druid as the first actor on the stage of human events in Great Britain. In this way the mind of the public has been for several generations prepared to accept unhesitatingly all the stuff that has been written about Druidical temples, sacrificial altars, sun-worship, and astronomical theories.

Barry, Hibbert, Scott, and Maculloch have each assailed the old Druidical theories with considerable learning and ability. Of Dr. Maculloch it has been remarked that he 'wielded the hammer of Thor with very signal success in aid of the demolition of the Druidical theory.' But those writers who have attempted to prove a Scandinavian origin for the Scotch circles have failed as much as the supporters of the old theory. That Norsemen entered chambered cairns for purposes of robbery is fully proved and recorded, and they have left marks and inscriptions therein. Would it not be reasonable to suppose that, had they made the circles themselves, they would have left carvings or inscriptions thereon ? Moreover, no tradition assigns these monuments to them. Stennis is the old Norse Steinsnes, that is, 'promontory of the stones,' showing that the old Norse rovers found them already there on arriving in Orkney. Some circles are clearly burial-places of the Bronze Age, and others may have been constructed by the people who built the big chambered cairns. This seems very probable.

## CHAPTER XII

### STONEHENGE NOT DRUIDICAL

> Thou noblest monument of Albion's isle!
> Whether by Merlin's aid from Scythia's shore
> To Amber's fatal plain Pendragon bore,
> Huge frame of giant-hands, the mighty pile
> To entomb his Britons slain by Hengist's guile;
> Or Druid priests, sprinkled with human gore,
> Taught mid thy massy maze their mystic lore;
> Or Danish chiefs, enrich'd with savage spoil,
> To Victory's idol vast, an unhewn shrine,
> Rear'd the rude heap; or, in thy hallowed round,
> Repose the kings of Brutus' genuine line;
> Or here those kings in solemn state were crowned.
> Studious to trace thy wondrous origin,
> We muse on many an ancient tale renown'd.
>                                         WARTON.

'EVERY kind of theory has been proposed and as regularly combated. And so it will be till the end of time. Each generation considers itself wiser than the preceding, and better able to explain those matters which to their fathers and grandfathers only appeared more difficult of explanation as they advanced in their inquiries. And thus it has come to pass that more books have been printed about the much-frequented Stonehenge than about all the megalithic structures, collectively, which the world contains; and the literature of this, the best known of them all, would fill the shelves of a small library.' So wrote Mr. William Long in his elaborate and careful paper [1] on Stonehenge, completing a work

---

[1] *Wilts Archæol. and Nat. History Mag.*, vol. xvi. (1876), p. 1, with maps and illustrations.

of research begun by Dr. Thurnam. And what he says is true, although it may be remarked that later researches among stone circles have thrown much light on the subject, and it is, therefore, quite possible that before very long the mystery of Stonehenge may be solved ! Already the antiquarians have prepared the way to an explanation, as was shown in the last chapter. It is so unique a monument, and so well known in these days of easy travelling, that the subject deserves to be treated in a separate chapter. Our object will be twofold—first, to show that it is *not a temple* ; and secondly, that it *was not erected by the Druids*—a conclusion for which the reader will be already prepared.[1]

Stonehenge is situated on Salisbury Plain, about two miles from Amesbury, and about eight from the city of Salisbury. Its very name speaks of a high antiquity, the termination being derived, according to Sir John Lubbock, from the Anglo-Saxon word 'ing' or 'eng,' a field. 'What more natural,' he says, 'than that a new race finding this magnificent ruin standing in solitary grandeur on Salisbury Plain, and able to learn nothing of its origin, should call it simply the *Place of Stones* !'

Other explanations have been given, but they are less satisfactory ; one is that it means 'hanging-stones' (Anglo-Saxon 'stane' used as an adjective, and some word resembling 'henge' signifying

---

[1] Amongst other sources of information are an account in Murray's *Handbook for Wilts, Dorset, and Somersetshire* ; *Rude Stone Monuments*, by Sir James Fergusson, whose theory of its post-Roman date cannot be accepted (the book is untrustworthy in other ways also) ; *Stonehenge : Plans, Description, and Theories* (a very careful and thorough study), by Professor W. M. Flinders Petrie, the celebrated Egyptologist ; *Anthrop. Institute*, vols. xi., xii., xv., xvii., xx., papers by Mr. A. L. Lewis, whose facts are useful, but his theories fanciful ; also Admiral F. S. Tremlett, vol. xiii.; *Archæologia*, see Index, under ' Stone Circles and Stonehenge ;' *Prehistoric Times*, Sir John Lubbock ; *Early Man in Britain*, Professor W. Boyd Dawkins. It is to be regretted that even recent writers speak of the Great Monument as a temple, for the phrase must surely be misleading. See also *The Encyclopædia Britannica*, and a good and sensible account of Stone Circles in *Chambers's Encyclopædia*. Also *Jottings on Stonehenge* (E. T. Stevens) ; *Guide to Old Sarum and Stonehenge* (Brown & Co., Salisbury), and *Archæological Review*, vol. ii. p. 312.

something suspended), but that idea would have been better expressed by 'hengestanes;' another suggestion is, that it connects the monument with the Saxon chief, Hengist, according to a tradition to which we will refer later on.

The monument *appears* to have consisted of two circles of upright stones, and two ellipses of a horseshoe shape occupying the central portion of a circular area having a diameter of about 300 feet, bounded by a bank and a ditch, as many other stone circles are, if not all. The entrance faced roughly to the north-east, and the road to it is still traced by banks of earth that marked the 'Avenue.' Probably in some connection with the monuments are traces of what appear to have been two 'alignments,' as at Avebury, on Dartmoor, and elsewhere. Some of the older archæologists saw in each of these a cursus, or hippodrome. But that view is now quite abandoned. Professor W. Flinders Petrie alone has observed parallel banks cutting across the Avenue.

Certain outlying stones are of some importance.

On approaching the circles by the Avenue one meets near the earth-ring an isolated stone called the 'Friar's Heel,' a block 16 feet 9 inches long, now in a leaning position, the top of which, to a person standing on the so-called 'Altar Stone' (inside the inner ellipse), exactly coincides with the line of the horizon, and at the summer solstice the sun rises directly over the top of the stone. This fact and the direction of the 'Avenue' are almost the sole foundation on which the several astronomical theories have been built up. The stone occupies a distinguished position in the legendary history of Stonehenge. According to one version, the devil, while busily engaged in erecting the stones, made the observation that no one would ever know how it was done. This was overheard by a friar who happened to be near watching the operation, and who incautiously replied in the Wiltshire dialect, 'That's more than thee can tell,' and fled for his life. Whereupon the devil caught up an odd stone, flung it after the friar, and hit him on the heel, but so holy was he that it did not hurt him in the least; the stone suffered instead, and was indented so that the mark of the friar's heel may be seen to this day.

On entering the earth-ring one sees a prostrate stone 21 feet in length, popularly known as the 'Slaughtering Stone.' It has been suggested that this stone once stood upright, but some writers reject the notion, because if it had been erected it would have prevented a person standing on the 'Altar Stone' from seeing the sun rise at the summer solstice over the 'Friar's Heel.' But then, who knows that the 'Altar Stone' was the place from which the sunrise was observed? and ought we not to be very careful lest we be prejudiced by preconceived ideas? Professor Flinders Petrie thinks the point of observation was intended to be from behind the central trilithon of the inner ellipse, or horse-shoe, as it is sometimes called. The 'Slaughtering Stone' has a row of holes worked across one corner, as if it had not been fully trimmed into its final shape. The holes were probably made with the intention of dividing the stone in that direction. But of this we shall have more to say presently. Whoever gave this absurd name to the stone evidently believed Stonehenge to have been a Druidical temple where human victims were sacrificed.

On going round the earth-circle, or rampart and ditch, one comes across two other stones (not erect) at about the same distance from it as the 'Slaughtering Stone,' and at opposite points of the circle. They are said to be unhewn, and it has been suggested that these, together with the 'Slaughtering Stone,' are remnants of a complete circle of standing stones within the rampart, as at Avebury, but the idea is not generally accepted. For our part, we would suggest that it is, at least, worthy of thorough and impartial examination;[1] and is it not just possible that the fact of the sun rising over the 'Friar's Heel' at the summer solstice is a coincidence? Once fill people's minds with astronomical theories and sun-worship in connection with stone circles, and one finds these ideas

---

[1] We are glad to see that Mr. Arthur Evans holds this view. See *Archæological Review*, vol. ii. p. 312.

not easily eradicated. Besides, has such a fact ever been noticed in connection with any other stone circles? Are there, or were there, stones in the avenues, or at the entrances to *other* circles, which can be shown to have served a similar purpose? We cannot help remarking that when people indulge in theorising about Stonehenge, they speak as if there were no other stone circles in the world!

One other fact should be recorded here before we pass on to the monument itself—namely, that there are, just within the earth-ring, and not very far from each of the two stones above mentioned, two round barrows. For plans of Stonehenge (which we are sorry not to be able to provide here) the reader should consult Murray's 'Handbook,' 'Rude Stone Monuments,' 'Early Man in Britain,' or Mr. Stevens's useful little book of 'Jottings on Stonehenge,' published at a popular price in Salisbury. Now with regard to these barrows it is important to notice their exact position. Most writers assert that they are partly intersected by the rampart, and are so represented in several plans, but Professor Flinders Petrie, who has made the most accurate plan by his own careful measurements, denies this, and says they are just beyond the rampart, and *not* intersected by it. Sir Richard Colt Hoare, who found an interment in one of them, says: 'From this [the supposed intersection] we may fairly infer that this sepulchral barrow existed on the plain, I will not venture to say before the construction of Stonehenge, but probably before the ditch was thrown up.' The small pits in the middle of each mound suggest that they have already been dug into and explored.

Coming now to the monument itself, it *appears* that the outer circle originally consisted of 30 upright stones fixed in the ground at intervals of about four feet (that being the space between one pair and the next), connected at the top by a continuous line of 30, or perhaps 28, imposts, forming a ring of stone at a height of about 16 feet above the ground. Only 17 uprights and 6 imposts are still standing, and, judging from these, they all must have stood about 12 feet 7 inches out of the ground, their average breadth being 6 feet, and their thickness 3 feet 6 inches. The imposts are about 10 feet long, 3 feet 6 inches wide, and 2 feet 8 inches thick. The diameter of this outer circle of united trilithons is about 100 feet, measuring to the inside surface of the stones. The blocks have all been hewn in some way, so that their sides are more or less parallel, and their corners squared. How that was done we must consider presently. They were all cleverly joined together. The uprights were cut so as to leave knobs or tenons, one at each end, which fitted into mortice-holes cut out in the under side of the imposts, an arrangement which is clearly a survival from an earlier time when woodwork was used instead of stone. Everywhere wood preceded stone for building purposes.

This outer circle consists entirely of a somewhat soft sandstone, known as 'Sarsen-stones,' the blocks of which are found in great abundance on the surface of the ground in North Wilts. They are fragments of Tertiary strata which once overlaid the whole of the chalk of Salisbury Plain, all the rest having been destroyed by the forces of 'denudation' that are constantly at work moulding and modifying the surface of the earth. Prof. Flinders Petrie points out that the number of stones in many other stone circles is also a number divisible by ten. 'Similarly, the circle at Winterbourne Abbas, Dorset, has 10 stones (one missing); the circles of Dawnsmaen, Boscawenoon, and Kenidjack, in the Land's End, are each of 20 stones. . . . Other cases, where the circle is less perfect, are less decisive; but the three circles of the Hurlers, near St. Cleer, are credited with having had 29, 29, and 27 stones, which may very possibly have been 30 originally. In Cumberland, the Eskdale circle was of 40 or 41 stones, the Swinside circle of 60 stones, and the Gunnerskeld circle, Westmoreland, 29 + or − 1, or spaces for stones which show apparently the same decimal division.'[1] The diameters of stone circles measured

[1] *Loc. cit.*, p. 15.

T

in feet are generally also decimal numbers, which is a curious fact, considering that our English foot cannot be the same as the old prehistoric unit of length (see Chapter XI. p. 259).

Some writers, *e.g.* Mr. E. T. Stevens, consider that the imposts of the outer circle were originally slightly dovetailed into each other at their extremities, but others do not support this idea ; in fact, certain writers appear to doubt whether their ends were in contact. Some of the published ground plans and restorations show the ends *not* touching. And here one cannot help remarking that the general effect of the monument would have been much more imposing if the imposts had formed a continuous stone ring, as in the restoration given by Stevens in his ' Jottings ' (p. 83). The photographs clearly bear out this idea in the case of the few outer trilithons left standing ; but Mr. William Long apparently does not think so (see the coloured restoration of the ground plan in his paper, p. 54).

Within this outer circle, and at an average distance of 9 feet from it, was an inner circle, composed of 30 (or 40) pillars of syenite, greenstone, &c., and other igneous rocks, quite foreign to the neighbourhood, each about 4 feet in height and 1 foot in breadth. But few of these are left standing. Mr. William Long says they are rude and irregular in shape, and apparently *un*wrought. Prof. Flinders Petrie, on the other hand, speaking of the stones generally, says : ' Nearly all of the stones had more or less dressing ; some dressed all over, others only on prominent parts, so as to give a regular figure at the end or edges, with faulty parts elsewhere.' These stones must have been brought from some place a long way off, but whether it was Cornwall, or Devonshire, Wales, Brittany, or Ireland, it is impossible to say. Geologists who have made a special study of Petrology say they might as well have come from South Wales as from Cornwall. The labour of bringing them must have been enormous, and it is highly probable that they were considered very sacred. Whether they stood each in front of one of the big sarsens of the outer circle or between each pair has not been determined.

Within this circle, again, was a group of enormous trilithons, quite separate from each other, and five in number, arranged in a horse-shoe form. But some authorities consider (on the evidence of a smaller stone with holes lying near the opening of the so-called Horseshoe) that this was really an ellipse formed of 7, or perhaps 10 trilithons, the missing ones being smaller. They were all certainly hewn, and the five were sarsen stones. These imposing structures rose progressively in height from north-east to south-west (not shown in Plate X.). The highest was just behind the so-called 'Altar Stone,' facing the entrance to the Horseshoe and the Avenue. Those that were furthest from the central group were smaller; only two are now perfect. The Duke of Buckingham, in the time of James I., in digging for treasure, appears to have caused the overthrow of the chief trilithon, so that one upright has fallen down and the impost, while the other upright is leaning over the 'Altar Stone,' supported by one of the inner greenstones. Another trilithon of the outer ellipse fell on January 3, 1797, on a rapid thaw succeeding a frost.

Next came the inner ellipse or horse-shoe of igneous rocks (called greenstones, bluestones, syenites, &c.), simple uprights varying in height from 6 to 8 feet, and set at intervals of about 5 or 6 feet or even 8 feet. They are, says Sir R. C. Hoare, much smoother and taller than those of the inner circle, and rather pyramidal in shape. But few of these are now left, so that it is hard to say what positions they occupied, or how many there were of them. Some restorations represent them as standing in five groups of three, each group being in front of one of the big trilithons of the outer ellipse. Others make the number 19 or 20 (the latter seems more probable, as there is a tendency to decimal numbers : see p. 273). They appear to have been trimmed. Within this inner ellipse lies the 'Altar Stone,' which is composed of a micaceous sandstone, also foreign to the district. Its length is 16 feet 2 inches. It lies askew, and *may* mark an interment. There is not a trace of the action of fire on it. It was broken in two by the fall of the impost of the great trilithon on to it.

On the left-hand side, as one enters the inner circle from the north-east, is a recumbent foreign stone, with two cavities worked on the surface which is now uppermost. Are they mortice-holes, and was the stone intended for the impost of another smaller trilithon? The question cannot at present be answered. We can see no particular objection to the idea that the two ellipses were originally, or were intended to be, complete.

With regard to the holes which have been made in the so-called Slaughter Stone, mentioned above, Mr. E. T. Stevens points out that they are oval in shape, and thinks the instrument used was a pick (*not* a chisel), and that it may have been of stone! This writer quotes cases in which flint tools have been used for cutting figures on hard granite.[1] But even if the stones were dressed and carved by means of flint tools, it would be unsafe to conclude that Stonehenge was erected in the Stone Age, for bronze tools may have been scarce, and possibly the flint ones worked better when properly managed. Moreover, it is unlikely that the builders of Stonehenge knew how to harden bronze as the Egyptians did; and we would suggest that the use of metal might have been forbidden for religious reasons in an age of stone worship.

The visitor to Stonehenge will observe mounds or tumuli ranged round. Within a radius of three miles there are as many as 300, a large number of which have been explored. They are mostly round barrows with interments by cremation, and in one or two cases objects of bronze have been found; but usually they contain only objects of flint, bones, and pottery, though glass and amber have been found, but nothing Roman. The relation of the barrows to Stonehenge has been much

---

[1] *Jottings on Stonehenge*, pp. 86, 99, 101.

discussed. Many of them must be distinctly later than the time when the monument was erected—namely, such as contain chips of the foreign stones as well as of the sarsens. This is a valuable little bit of evidence. The general grouping of the barrows does not show any particular reference to Stonehenge. There is, however, a gap in the line of barrows in the direction of the east branch of the 'Avenue,' which *appears* to show that the barrows were made afterwards, though it is not absolute proof. A cluster, half a mile to the north, is called the 'Seven Barrows,' and adjoining it is the western end of the so-called 'Cursus.' This long enclosure is marked out by banks of earth along the plain, east and west, to a distance of more than a mile and a half. At one end is a high mound. Of course, no serious archæologist now believes that this was a race-course. A little to the north is the second and smaller one, also with a bank at the west end. The 'Avenue' extends from Stonehenge more than 1,700 feet in a straight line towards the north-east.

Prof. Flinders Petrie thinks there are several evidences of incompleteness, and therefore that the monument never was finished. The two circles are not concentric, and many writers consider that the inner circle of igneous rocks, and the inner ellipse, also of stones foreign to the district, belong to an older monument, the great horseshoe and the great outer circle, both composed of trilithons, being added afterwards. But why this theory has been so widely accepted we fail to understand. For how could the biggest imposts (some 25 feet long) have been dragged through the supposed older circle, and then set up? Would it not have been a very awkward job, unless these stones were moved? and that is unlikely, because they were most probably sacred. Also the stones of the inner ellipse would get in the way, especially if banks of earth were used in raising the stones. There is a tradition that the foreign stones came from Ireland, being brought by the aid of Merlin, the magician. They *may* have been sacred stones long before Stonehenge was set up, but since some of them are trimmed, we can see no particular reason against attributing one period only to the work of erection.

Bearing in mind the alignments of Brittany and Dartmoor, and their connection both with barrows and stone circles, *and* the facts already brought forward in the last chapter showing that so very many such circles have been proved to be sepulchral, the memorial character of menhirs, *and* the clear analogy between a dolmen surrounded by a ring of upright stones and a stone circle—to say nothing of the more ancient 'long barrow,' with its surrounding stones—we think that the word 'temple' is an inappropriate one for a monument, even if it were a sacred place. The very idea of worship in the ordinary sense had not yet been developed. Sacrifices were the only form of public worship then in vogue—say, in the early days of the so-called Bronze Age. The stone circle, probably, was the prototype of the temple, for tombs preceded temples. Again, the trilithon is very suggestive of the dolmens and their megalithic passages, formed of upright stones with an impost for roof.

There has been a great deal of speculation as to the manner in which the huge blocks were transported and raised, but the matter is comparatively simple after all, even without machinery. In old days people were in no hurry, and kings or chiefs could command the labour of large bodies of men. They had no pulleys (even the clever Egyptians were without them), only ropes, levers, crowbars, and rollers, but wonders can be done with these. Sir James Fergusson says: 'And anyone who has seen with what facility Chinese coolies carry about monolithic pillars ten feet and twelve feet long, and thick in proportion, will not wonder that twenty or thirty men should transport these from the head of Southampton Water to Stonehenge.'[1] With regard to their erection, the usual theory is that the big stones were dragged over rollers along inclined mounds. Now there are some obvious drawbacks to this method; it is very cumbrous, and involves

[1] *Rude Stone Monuments*, p. 95.

great waste of labour. Fancy making a separate embankment of earth, or earth and stones, for the erection of each of the great stones of the trilithons ! As before mentioned, it is fatal to the idea of the 'Sarsen-stones' being added afterwards. Feeling dissatisfied with this idea, we appealed to Mr. C. H. Read, of the Department of Antiquities in the British Museum (who has just succeeded Sir Wollaston Franks as head of the department), and afterwards to Prof. Flinders Petrie, both of whom suggested that the stones were simply wedged up with levers. The professor kindly drew us a diagram to illustrate the idea, saying that it is a method even now largely in use, and one which he often uses when at work among the ruins of Egypt. Once wedge up a block of stone, and put something under it near the middle, and it is wonderful what you can do with it. The reader will find this idea worked out in Plate X. Probably wedges of wood and then blocks of stone were placed under the monoliths as they were slowly and laboriously wedged up with long wooden crowbars. After a time, the men would require to raise their position. This could easily be done by means of stones or timber, and so a wall of neatly-trimmed stones was gradually built up under each impost in order to raise it to the required height. It could then easily be slipped along on rollers till it rested accurately over the uprights. We are greatly obliged for this excellent suggestion, especially as a long embankment would spoil the composition of the picture. A stone axe is seen lying on a big block in the foreground. An overseer is looking on. The men's trousers have, perhaps, rather a modern look, but we have good evidence that such garments are of very ancient origin : we see them worn by the Indians of North America, and we know from the evidence of monuments that ancient Gauls and Germans wore them.

The 'weathering' of the stones has been considerable, and that speaks of high antiquity.

It is not worth while to trouble the reader with details of the absurdities of the various astronomical theories which from time to time have been put forward, but we may mention that Stonehenge has been called :—

1. A temple of the Sun.
2. A temple of Serpent worship.
3. A shrine of Buddha.
4. A planetarium, or astronomical model of the planets.
5. A calendar in stone for the measurement of the solar year.
6. A gigantic gallows on which defeated British leaders were hung in honour of the Saxon god Woden.
7. A memorial set up by Aurelius to commemorate the British nobles treacherously slain by Hengist the Saxon, at a banquet.

The last explanation we owe to Geoffrey of Monmouth (twelfth century) and the Welsh bards, but it can only be treated as mythical.

Stonehenge is not mentioned by any Roman writer, which is hardly surprising (nor do they notice Carnac, in Brittany). We have no mention of it by such old writers as Gildas, Nennius, or the Venerable Bede, but in the twelfth century it is referred to by Henry of Huntingdon as one of the four wonders of England, the other three being merely natural phenomena.

Among the people to whom its construction has been attributed are—the Druids, the Phœnicians, the Belgæ, the Saxons, and the Danes !

We wish that more digging and excavating could be done here, under proper superintendence, and we confess to a lurking suspicion that interments might be found in connection with some of the stones. The entire absence of inscriptions or markings of any kind (except such as have been made by visitors) is in itself one proof among many of a high antiquity. Who made the monument?—*i.e.* what race of people, is a question which cannot at present be answered positively. But it would appear to be connected with the people who constructed the barrows, either the long ones or the round ones. Was it the long-headed Iberians who preceded the Kelts? Possibly so. There is nothing to connect it with the Druids, who worshipped in groves, nor with any of the Kelts, whose temples were small wooden square-shaped structures.

Col. Forbes Leslie, in his 'Early Races of Scotland,' points out that in India, in the Dekkan, are certain circular monolithic 'temples' still used for worship, in which the relative positions of certain stones agree in a remarkable manner with those at Stonehenge; thus, the monolith known as the 'Friar's Heel' (also as the Gnomon, an astronomical term), which stands outside the circle, would have its counterpart in these temples. We should like to know whether this observation has been confirmed by others. If true, it is of great importance, but we have been unable to find notices of any other cases. The above writer speaks of a Hindoo fane in Western India, on the tableland above the Ghauts, in which a cock had recently been sacrificed to Betal, and says it consisted of twenty-three small stones, placed in a circular form at equal distances; one to the east was moved twelve feet back, three smaller stones were outside, and to the south-west a single stone, but no opening. But the question arises, How old were these circles? Later generations of men might use them for any purpose they pleased, quite independently of the original purpose for which they were set up.

The Keltic people of Scotland used stone circles as meeting-places, as well as for public worship; but they themselves were quite in ignorance of the purpose, or purposes, for which these monuments were erected; the same argument, then, applies to customs and practices of modern Hindoos or others. According to Colonel Meadows Taylor, large rocks, with circles round them, are used as places of sacrifice by Indian shepherds. Maurice ('Indian Antiquities,' p. 158) says all ancient temples of the Sun and Vesta, or elementary fire, were circular; the adytum in which the sacred fire blazed was constantly of an egg shape. The natives, near the first cataract of the Nile, are said to worship in circles of stone four or five feet high, on the tops of hills ; Pausanias also refers to circles of great stones in which the mysterious rites of the earth-goddess Demeter were performed. Mr. W. Long, in his paper above referred to, quotes a remarkable account given by Mr. W. G. Palgrave of a structure very similar to Stonehenge, which he found in Arabia,[1] from which we extract the following :—

'But immediately before us stood a more remarkable monument, one that fixed the attention and wonder even of our Arab companions themselves; for hardly had we descended the narrow path where it winds from ledge to ledge down to the bottom, when we saw before us several huge stones, like enormous boulders, placed endways perpendicularly on the soil, while some of them yet upheld similar masses laid transversely over their summit. They were arranged in a curve, once forming part, it would appear, of a large circle, and many other like fragments lay rolled on the ground at a moderate distance ; the number of them still upright was, to speak from memory, eight or nine. Two, at about ten or twelve feet apart one from the other, and resembling huge gate-posts, yet bore their horizontal lintel, a long block laid across them ; a few were deprived of their upper traverse; the rest supported each its head-piece in defiance of time and of the more destructive efforts of man . . . . The people of the country attribute their erection to Darim, by his own hands, too, seeing that he was a giant, also for some magical ceremony, since he was a magician ; pointing towards Rass, our companions affirmed that a second and similar stone circle, also of gigantic dimensions, existed there ; and, lastly, they mentioned a third towards the south-west. That the object of these strange constructions was in some measure religious seems to me hardly doubtful, and, if the learned conjectures that would discover a planetary symbolism in Stonehenge and Carnac have any real foundation, this Arabian monu-

[1] *Narrative of a Year's Journey through Central and Eastern Arabia*, 1862-63, by W. G. Palgrave, vol. i. p. 250 (1865).

ment, erected in the land where the heavenly bodies are known to have been once venerated by the inhabitants, may make a like claim ; in fact, there is little difference between the stone wonder of Kaseem and that of Wiltshire, except that the one is in Arabia, the other, the more perfect, in England.'

Mr. John Henry Parker, of Oxford, author of the 'Glossary of Architecture,' at a meeting of the members of the Wilts Archæological Society in 1865, said that in Oriental language a circle of stones was called a Gilgal, and in Scripture there was every reason to believe that such a place was a circle of stones. A Gilgal, he thought, was a temple where holy rites were celebrated, where the army met together, and was also used for a place of burial for the chieftains. . . They might, therefore, call it a burial-place, or a House of Commons. Let it be noticed that he uses the word temple in a modified sense merely as a sacred place. Mr. Parker's conclusion seems eminently reasonable, especially in ignoring the elaborate astronomical theories of some writers. But even he seems to connect it with the Keltic people, as was the fashion twenty or thirty years ago.

Prof. Sven Nilsson, the venerable Swedish antiquary, read a paper before the Ethnological Society of London in July 1868, in which he compared Stonehenge to a similar temple at Kirik, in Sweden, and with solar temples in Syria and Phœnicia. He considered it to have belonged to that period when bronze was used for weapons and implements in Western Europe, and he believed it to have been devoted to solar worship.

Mr. E. H. Palmer, in his 'Desert of the Exodus,' speaks of 'huge stone circles in the neighbourhood of Mount Sinai, some of them measuring 100 feet in diameter, having a cist in the centre covered with a heap of large boulders. These are nearly identical in construction with the so-called "Druidical circles" of Britain. In the cists we found human skeletons, the great antiquity of which was proved not only by the decayed state of the bones, but by the fact that the bodies had in every case been doubled up and buried in such a position that the head and knees met. There are also small open enclosures in the circles, in which burnt earth and charcoal were found.' This discovery clearly shows that all 'cromlechs' were *not* covered with earth.

If the 'little-folk,' or dwarfs, built the chambered cairns, as seems proved, why should they not have built Stonehenge, together with some other stone circles, *e.g.* Calernish? They were said to be great builders. This idea was suggested to the writer by the analogies with the chambered cairns and dolmens. The trilithon itself is evidently derived from the covered ways of dolmens. Everybody has his theory about Stonehenge. We only offer this suggestion with the object of stimulating further discussion of the subject in the light of recent discoveries and theories about the pre-Keltic dwarfs.

In conclusion, we venture to offer the following observations and arguments :—

1. Stonehenge must be studied in connection with stone monuments generally all over the world and in every age. It is not a unique phenomenon to be judged entirely by itself.

2. Stone circles are clearly connected with 'avenues' and 'alignments' of stone.

3. Upright 'pillar-stones,' 'menhirs,' or obelisks, are known to have been sacred, and often objects of worship. The references to such in the Old Testament are numerous. And even at the present time stones can be seen in India which the natives have placed to represent their gods.

4. Stonehenge, being partly composed of such upright stones, may have been set up with a religious object.

5. But it does not follow that we have here a 'temple.' That word, as already remarked, is rather misleading.

6. Some of the barrows appear to be later than Stonehenge —viz. those containing chips of the igneous rocks foreign to the district. It is therefore possible that *all* of them were later. The

gap in the 'Kings' Barrows,' intersected by the Avenue, points in the same direction.

7. Many stone circles are proved to be sepulchral—*e.g.* in England, Scotland, Scandinavia, Mount Sinai. Avebury is so.

8. Few writers on Stonehenge, if any, have noticed the important fact with regard to certain stone circles, that they sometimes occur *in groups*; and cases are known where they even intersect one another (*e.g.* Boscawen), or make a pattern of contiguous circles. Is it likely, then, that these could be temples?

9. The fact that interment in stone circles took place in Scandinavia and Algeria as late as the Iron Age is an important fact, and probably a survival of a very ancient practice.

10. The absence of all reliable tradition points to a remote antiquity. The story of Ambrosius and the treachery of Hengist probably refers to Amesbury (Ambresbury). Geoffrey of Monmouth and the Welsh bards are unreliable.

11. No Roman writer mentions Stonehenge. If Druids had worshipped in stone circles, we should surely have been told so; whereas we know they worshipped in groves.

12. Stone circles were not erected in Britain or Ireland by Scandinavians, but were here before the Norsemen came. The veneration in which these structures were held also denotes a high antiquity.

13. They were used as meeting-places and courts of justice both by Kelts and Scandinavians, and possibly such use was merely a continuation of the ancient use.

14. The majority of them in Great Britain appear to have had their entrance facing the north-east or east, but some the south.

15. The obvious analogy with dolmens and their surrounding circle of stones; their megalithic passages and general ground plan bearing a striking resemblance to some circles and avenues —*e.g.* Calernish.

16. But it would not be safe to conclude that *all* stone circles are places of burial

17. If they were raised by the people who made the round barrows, that would be an argument against their sepulchral use, for it is unlikely that barrows and stone circles were *both* used at the same time, and by the same people, for burying their dead.

18. But why should they not belong to the people of the long barrows and the chambered cairns, as suggested by the Calernish monument? Number 6 is in favour of this view, as all the barrows near are round ones. Again, Stonehenge may be purely memorial, as tradition says.

# INDEX

ABBEVILLE, discoveries at, 17
Abraham and the Hittites, 202
Abri, the, of La Laugerie, 67
Ainos of Japan, the, 68, 225
Alignments, 250, 251, 284
'Altar Stone,' the, at Stonehenge, 270
Amulets, 215
Ancient sages, 2, 3, 6
Andalusia, antiquities of, 237
Andaman islanders, 77
Animals contemporary with Palæolithic man, 33, 55, 66; Neolithic man, 183
'Animism,' 83, 203
Antiquity of man, 20, 27, chapters vi. and vii.
Anvil stones, 32
Apes, Miocene, 128
Apsides, revolution of the, 132
Architecture, evolution of, 10, 41, 200, 201
Aristotle, 3, 4
Arnold, Matthew, 233
Arrow-head in vertebra of horse, 64; at Spy, 25
Arrows, magical, 214
Astronomical theory of Ice Age, 116-127
Atmosphere, changes in the, 109
Austen, Major Godwin-, 252, 253
Avebury, 250, 262, 263, 285
Awl, of flint, 67
Axis of earth's rotation, 113

BABYLONIAN tablets, 6
Bacon, Sir Francis, 167
Ball, Sir Robert, 117
Barrows (tumuli), Greek traditions, 196; Chinese custom, 196; Homer's account of Hector's funeral, 197; the stone pillar on, 197; King Gorm's tumulus, 199; Harald's tumulus, 199; Lycian tombs, 200; origin of architecture, 200; Siberian yourt, 201; classification of British, 205; long barrows, 206; Uley barrow, 206; upright stones on barrows, 206; Homeric references to, 206; in France, 210
Basques. *See* Iberians
Bâtons de commandement, 70
Bear, cave-, at Duruthy, 65; brown bear, cave-bear, 66
Black, Mr. W. G., on Finns of Schleswig, 224
Blackmore Museum, Salisbury, 173; Dr. Blackmore, 25
Blind, Dr. Karl, on Finns, 230
Bones, split, 64
Bonney, Professor T. G., on the ice-sheet, 87; on the temperature of the Glacial period, 125
*Bos primigenius*, 182
'Botanic Garden,' the, 37
Boucher de Perthes, M., 17, 18
Boulder clay, 22, 88, 89, 90
Boulders, erratic, 84, 92, 94, 96; Scandinavian, on English coast, 97, 98; in Alps, 101
— pedestal, 147
Bourgeois, the Abbé, 156, 157
Bows and arrows, 54
Boyne, Brugh of, 221
Brogar, ring of, 264

Bronze Age, duration of, 244
Buckland, Dean, 42, 45, 46, 49, 93, 95
Burmah supposed Tertiary man in, 154

CADDINGTON, Palæolithic site at, 30
Cairns, 207; at New Grange, 208
Callernish monument, 261, 263
Campbell, Mr. J. J., of Islay, on the dwarfs, 224; on Iberians, 235
Camus stones, 248
Canstadt skull, 75
Carnac, alignments at, 251
Cat stones, 248
Cataclysm, 93
Caves, deposits in, 17; list of British, 45; sacred caves in Greece, 40; dragons in, 40; formation of, 44; Bosco's den, 52; Kirkdale Cavern, 47; Kent's Cavern, 54, 57; cave man (Palæolithic), his weapons, 54; King Arthur's Cave, 59; Cae Gwyn Cave, 60; Victoria Cave, 61; supposed pre-Glacial deposits in, 60; caves in the Pyrenees, 64; in other countries, 63; in Sicily, 104
Celestial sphere in Rome. *See* Eudoxus
'Challenger' expedition, 114
Chambers, Mr. Robert, on Picts, 226
Charpentier, 94, 95
Child Roland, 225
China, loess of, 103
Christy and Lartet, their work in South France, 62, 68
Chronological scheme of Worsaae, 244
Chronology of Archbishop Ussher, 17
Civilisation, beginning of, 129
Civilised man, 16
Clachan (Gaelic), 259
Clarke, S. B., on Khasia Hill tribes, 257
Climate of Palæolithic period, 55, 57, 58. 60, 66, and chapters iv. and v.
Clothing of the reindeer hunters, 69
Clunie Castle, hill of, 220
Coal, in lat. 81° N., 108
'Contorted drift,' the, 22, 31
'Contracted position,' the, 209
Cooking, origin of, 78

Cooling of the earth, 109
Copper in dolmens, 245
Crannoges, 170, 175, 189, 190
Croll, Dr., 20, 86, 96, 98, 115, 116, 118, 121, 122, 145
Cro-Magnon, skulls of, 60, 74, 75
Cromlech, 241
Cup-markings, 243
Cuvier, 42
Cyclopean architecture, 227

DANES, 228
Darwin, on evolution, 3, 4; on the place of man's first appearance, 15; on 'the great submergence,' 91; on Glacial theory, 93; on Fuegians, 195
— Erasmus, 37
— George, on changes in earth's axis, 113; on astronomical theory of Glacial period, 117
Dawkins, Professor Boyd, on Galley Hill skeleton, 28; his book on cave-hunting, 45; on Kirkdale Cavern, 50; on migration of animals in Palæolithic period, 58; on the cave men of Périgord, 64, 73; on arrow-straighteners, 71; on the Neolithic homestead, 236
'Devil's Arrows,' 257
Diluvial theory, the, 45, 46
Dolmens, 241; builders of the, 244; etymology, 241; minerals in, 245; carvings in, 245; distribution of, 243
Dordogne, caves of the, 63
Drifts of the plateau, 22
Drinking cups in barrows, 213
Druidical theories about stone circles, 265
'Druids' Altars,' so-called, 241, 243
Dubois, Dr., 16; on the Java skull, 159-163
Dürnten beds, the, 151
Dwarfs, or little folk, 9, 40, 214, 216, 217; magical arts of, 217; the Sagas on, 217; J. J. Campbell on, 224; in kayaks, 229; possible builders of Stonehenge, 284 (according to the author)
Dwelling, the oldest, 40

ECCENTRICITY of earth's orbit, 118
Ecliptic, obliquity of, 113
Egyptian gods, 223; traditions, 238; new discovery in Egypt, 239
Elephants, fossil, 42
Elevation as a cause of Ice Age, 125
'Elf-shots,' 215
Elves, 9
Engravings by reindeer hunters, 65; list of animals engraved, 65
Erd-houses, 227
Eudoxus, his celestial sphere, 131
Evans, Sir John, 21; on Galley Hill skeletons, 28; on periods of Palæolithic age, 72
Eve, creation of, 7
Evolution, some new law of, 4; implied in Genesis, 6, 7, 10; perceived by Greek philosophers, 37
Exploration of caves, more needed, 68

FAIRIES, or little folks, the, 9, 40, 84
Falconer, Dr. Keith, 18; on antiquity of man, 154
Faroe Islands, 98
Fauna of Palæolithic period, the three groups, 56; of river-drifts, 33; supposed migrations in Palæolithic times, 57; of France in Palæolithic period, 66
Fergusson on Runic inscriptions, 208; on Stonehenge, 269
Fire, how obtained, 78
Flint implements of the plateaux, 23; uses of, 77; flint knife, how made, 76
Flood, the Noachian, 45, 47, 93, 105; floods in Glacial period, 100
'Floors,' Palæolithic, 30, 31
Fox-glove, probable etymology, 219
'Friar's heel,' the, 270
Funeral feasts, in connection with barrows, 211

GAELS, and the Tuatha Da Dannan, 223
Gairloch, the Tawny Hill of, 221
Gamme. *See* Laplander
Gap between Palæolithic and Neolithic periods, 72

Geikie, Sir Archibald, on legends about erratic boulders, 85; on boulder clay, 96
—— Professor James, 55; on fauna of older cave deposits, 55; on polar ice-cap, 86, 97, 98; on interglacial periods, 100, 125-127
Genesis, account of creation in, 6, 7
Geography, changes in, 21, 24
Glaciers, Alpine, 94, 149
Goethe on an Ice Age, 94
Gold in barrows, 210
Gomme, Mr. G. L., 225
Gravels, river, 10, 18, 19, 22, 144
Great Ice Age, 86, and all chapter iv.
Greenland, Danish report on, 97; Miocene flora in, 107, 108; ice-flow of, 149
Greenwell, Canon, 204; on cremation and inhumation in Yorkshire, 209, 210; on funeral sacrificial feasts, 212
Grinnell Land, Miocene flora of, 108
Gulf Stream, effect of, 122

HABITATIONS of Palæolithic man, 70
Hackpen, 263
Hall, Sir James, 92
Harpoons, 53, 54, 69
Harrison, Mr. B., finds plateau implements, 23
Hawk stone, 248
Hearths, Palæolithic, 73
Hengist, 270
Herodotus on the Nile, 136; on Pæonians, 167
Herschel, Sir John, theorem of, 119, 121
Hicks, Dr. Henry, on Cae Gwyn Cave, 60
Hill, Rev. E., on changes in earth's axis, 115
Hippocrates, on people of Phasis, 168
Hirdmane Stein, the, 259
History, two kinds of, 8
Hoar stones, 248
Hoare, Sir Richard Colt, on Stonehenge, 262
'Hogboy,' 208
Homer on Greek burials, 195, 197
Horace, quoted, 8

U

Horses, fossil, enormous number of, at Solutré, 64, 65
Howorth, Sir Henry, preface by, v.; on plateau implements, 24; on contorted drift, 31; on use of bow and arrow by Palæolithic man, 54; on supposed pre-Glacial deposits in Cae Gwyn Cave, 60; on polar ice-cap, 86, 87; on Schimper's discovery, 94; on Scandinavian boulders, 98; on diluvial theories, 103; on the great gap, 106; on Mr. Peck's theory of origin of signs of the Zodiac, 130; on age of Niagara Falls, 143; on Duernten beds, 151
Hughes, Professor T. McKenny, on Cae Gwyn Cave, 80; on supposed cuts on bones, 158; on pedestal boulders, 148; on St. Prest bones, 157
Huts, 70; of Germani, 176; of lake-dwellers, 176
Huxley, T. H., on Spy man, 25; on man and apes, 159
Hyænas in caves, 47-50

IBERIANS, 207, 235, 246
'Eis-zeit,' Karl Schimper, 95
Ice-cap, the, 86, 101, 105, and all chapter iv.
Iceland, stone utensils in, 76
Ice-sheet, the British, 96
Incense cups from barrows, 213
Interglacial deposits, 100
— periods of Professor J. Geikie, 100
Ireland, chambered cairns in, 208
Isaiah on stone worship, 259

JACOBS, Mr., English fairy tales, 9
Jade, in lake dwellings, 177
Java, skull, &c., from, 16, 159-163

KELVIN, Lord, on the earth's cooling, 109; on changes in earth's axis, 113-115; on geological time, 124
Kennet, West, barrow at, 206
Khamsin, the, 134
Khasia hill tribes, their standing stones, 251-255; how they transport stones, 254; religion of, 255

Kidd, Dr. Benjamin, on savages, 36, 80, 81
King's stone, the, 249
Kingussie, Cave of Raits, 227; meeting at the Standing Stones, 249
Kirkdale Cavern, 47-49
Kit's Coty House, 243
Kitchenmiddens, hearth-stones of, 192; molluscs, 193; weapons, 193; age of, 193; fauna of, 194
Knife of flint, how made, 76

LAING, Mr. Samuel, 156, 158
Lake-dwellings, localities of, 169-171; literature, 170, 190; in Britain, 170; periods of, 172; fauna of, 183; food of inhabitants, 183, 184; people of, 184, 185; pure copper found in, 185; Marin, 186; human bones, 187; interments, 188; crannoges in Britain, 189
La Laugerie, 67; crushed man of, 75
Lamarck, 3
Laplander's hut, or gamme, 202
Lapps, 41; once in Denmark, 215; Finns and Lapps, 230-235; same as 'Little Folk,' 208
'Lapp-shots,' 215
Larmer grounds, the, of Gen. Pitt-Rivers, 205
Law of progress, 4
'Law-Tings,' 259
Leslie, Colonel Forbes, 281
Lewis, Mr. A. L., on transport of big stones, 257
Lia Fail, the, 248
Lignite in Grinnell Land, 108
'Little Folk,' the, 40, 223; Tschuds in Russia, 223
Loess, the, 101, 102; human remains in, 75, 102; how formed, 102-103
Loffoden Islands, the, 97
Long, Mr. William, on Stonehenge, 268, 274, 282
Lucretius on stages of culture, 17
Lyell, on absence of human bones in river gravels, 26; on Glacial submergence, 91, 93; on uniformity, 101, 152; on cause of an Ice Age, 125

# INDEX

Lytton, Lord, translation of German poem, 204

MACBETH and the Witch Stone, 248
MacEnery, Rev., on Kent's Cavern, 18
Macpherson, Mr. A., of Kingussie, 249
MacRitchie, writings of, 221, 230; on Santa Claus, 225; on Picts, 226
Maeshowe, 207
Mammoth, 33, 42, 56, 59, 62, 99, 102
Man, antiquity of, 128, and all chapter vi.; Spy man, 25, 26. *See* Skeletons
Maories, intellectual capacity of, 37
Marsh, Professor, on the Java skull, 161
Max Müller, Professor, on religion of savages, 81
Ménec, 250
Menhirs in France, 249, 250, 284
Mentone, skeletons at, 75
Mermaids, 231
Meteorites in space, 110
Missionaries on religion of savages, 81
Mississippi, age of, 141
Monsters, bones of, in caves, 84
Morbihan, menhirs at, 250
Mundane egg, the Greek, 3, 7
Mykene, 220
Mythical school, the, 230

NATURAL selection, 4
Nebular theory, the, 109
Necklaces of stone beads, 33
Needles of bone, from caves, 67
New Grange, cairn at, 243
New Zealand, conditions of climate, 88
Newton, Mr. G. T., on Galley Hill skeleton, 88
Niagara, age of, 142
Nightmare, the Glacial, 86-88
Nile, age of the, 140; rising of, 136
Nilsson, Professor Sven, his theory of origin of tombs and houses, 202; his book on primitive inhabitants of Scandinavia, 215; on dwarfs, 218; on Stonehenge, 283
'Nine Maidens,' the, 251

Noetling, Dr., 154
Norsemen, 208
North American Indians, 76

OCEAN-BASINS, supposed permanence of, 114
Oldham, Mr. R., on Dr. Noetling's discovery, 155
— Thomas, Dr., 254
Orkney Islands, 98
Osiris, legend of, 233
Ossuaries, 211

PAINT-POTS, of horn, in Dordogne, 64, 67
Palæolithic man, date of, 60, 116, 244
Palgrave, W. G., on stone circles in Arabia, 282
Palmer, Mr. E. H., on stone circles in Sinai, 242, 283
Parallelitha, 251
Parker, Mr. John Henry, on Stonehenge, 283
Peat, rate of growth of, 151
Peck, Mr. W., his theory of origin of signs of the Zodiac, 129-138
Pengelly, Mr. W., 54
Périgord, the caves of, 62-83
Periods of prehistoric time, 11; of older Stone Age (De Mortillet), 71; probable duration of periods, 244 (Worsaae)
Petrie, Professor Flinders, on Mr. Peck's theory, 130; on Stonehenge, 269, 270, 274, 277, 279
Picts, 227
Pile dwellings. *See* Lake dwellings
Piper of Hamelin, the, 220
Pliny, on Frisian lake-dwellers, 168
Pluvial period, supposed, 145
Pogamogan, or bâton, 70
Pottery, absence of, in Palæolithic period, 70; in lake dwellings, 180; in barrows, 210
Precession of the equinoxes, 120
Prestwich, Sir Joseph, on river gravels, 22; on plateau gravels, 23; on loess, 103; on the Flood, 103; on erosion of river valleys, 144; on pedestal boulders, 148; on Thenay implements, 157

Primeval man, Mr. Worthington G. Smith on, 33-36
'Principle of Perfection,' the, 4
Pryse, Rev. J., on Khasia hill tribes, 254

QUATREFAGES, M. DE, 157

RAMESES II., statue of, 140
Rath, or hill-fort, 249
Read, Mr. C. H., on Stonehenge, 279
Record, nature of the, 8, 15
Reindeer hunters in South France, chapter iii.; literature of the subject, 62; herds of reindeer, 65; engraving of, 69
'Reliquiæ Diluvianæ,' 45, 46
River valleys, cutting down of, 145
Rivers, Pitt-, Gen., his museum at Rushmore, 205
Roches moutonneés, 89, 92, 125
Rock shelters in South France, 63
Rock tombs, 41
Rope bridles on Palæolithic engravings of horses, 71
Round-headed race of Bronze Age, 211
Rude stone monuments, distribution of, 243, 246

SAGAS, the, 214
Sahara, the, 21
Salisbury Museum, the, 25
Samoa, wives of, in battle, 79
Samoyedes, 202
Santa Claus, 225
Sarcen stones, 273
Saussure, De, 92
Sayce, Professor, on constellation Capricornus, 136
Schimper, Karl, discovers the Ice Age, 94
Schmerling, Dr., 18, 42
Scone, the coronation stone at, 248
Secondary interments, 211
Separate creations, 6
Shelters, rock, 70
Siberia, 124
Silbury Hill, 263

Sinclair, Dr. Robert, on Finns, 232, 235
Sivalik Hills, 154
Skeletons, human, at Spy, 26; at Galley Hill, 27; at Cro-Magnon, 68; at Mentone, 75; in Java, 159
Skertchley, Mr., on pre-Glacial man, 99, 147
Skins, for clothing, 29, 67
Skulls, of Spy, 26, 75; of Galley Hill, 27; of Neanderthal, 28, 72, 75, 161; Suffolk, 28; Cro-Magnon, 68; Mentone, 75
'Slaughter Stone,' the, Stonehenge, 271
Smith, Mr. Worthington G., on flint implements, 18; on fragment of human skull in Suffolk, 29; on Palæolithic man, 32, 33 35
Somme Valley, discoveries in, 17
Spindle-whorls, 179
Spitzbergen, Triassic beds in, 108
Spurrell, Mr., finds Palæolithic sites, 30
Spy, skeletons at, 25; Huxley on, 26
Stakes made by Palæolithic man, 33
Stalactites in caves, 44
Stalagmite, rate of growth of, 151
Standing stones, 246-250
Stars, variable, 111
Steinbergs, 174
Stevens, Mr. E. T., 274, 276
Stoke Newington, Palæolithic floor at, 30
Stone Age, the, 16, 19. *See* Periods
— circles in Mount Sinai, 242; remarks on, 258 267; in Scotland, 260; at Callernish, 263; Stennis, 264; numbers of stones in, 273; as law courts, 258; dimensions of, 259
— hammers, boring of, 178, 180
Stonehenge, chapter xii.; origin of the blue stones, 274; the avenue, 270, 277; the cursus (so-called), 270; on transport of big stones, 257; how raised, 279; theories of, 280; opinions of J. H. Parker and Professor Sven Nilsson, 283; the author's theory, 284
St. Prest, bones from, with supposed cuts, 157
String bridge, the, 256

## INDEX

Stukeley's Druidical theory, 243
Submergence, the great, 91
Sun, the proper motion of, 110
Swedenborg on erratics, 92

TACITUS on Germani, 79, 227; on cave-dwellers of the Red Sea, 80
Tanist stones, 248
Tannhäuser, the Venusberg, 220
Tasmanians, 36, 83
Taylor, Canon Isaac, on Turanian race, 203
— Captain Meadows, on Indian dolmens, 242, 281
Temple, the evolution of, 41
Temples, circular, in India, 281
Tents, possibly used by Palæolithic man, 70
Terraces, used for cultivation, 226
Tertiary man, supposed cases of, 154-163
Thames, gravels of the, 19; former course of, 20
Thank-offering of Khasia hill tribes, 257
Thayingen, engraving of reindeer from, 69
Thenay, implements from, 157. *See* Tertiary Man
Thorgils, Saga of, 228
Thurnam, his work on British barrows, 205, 211; on stone circles, 260
Tiddeman, Mr., on Victoria Cave, 61
Tiger, sabre-toothed (*Machairodus*), 53, 54, 60
Tombs, Turanian, 204
Transport of big stones. *See* Stonehenge

Tree of life, the, 3, 5
Tschuds, or dwarfs, in Russia, 223
Tuatha Da Danann, 221
Tyndall, Professor, on production of Glacial period, 112

UNDERGROUND houses. *See* Erdhouses
Urus, the, 51
Ussher, Archbishop, 17

VENETZ, 94
Vèzère, rock shelters of the, 63
Vikings, 98; derivation of the word, 245
Virgil on burial of Dercennus, 196

WELLS, Mr., his 'Time Machine,' 12, 129
White Horse, Vale of the, 251
Wilson, Archdeacon, 155
Witches' stanes, 248
Wookey Hole Cave, 50; fauna of, 52
Worsaae, chronological scheme of, 244
Worship of stones, 247

YOURT, 201

ZODIACAL constellations, origin of the, 128-138

# LITERATURE

*The more recent books are indicated by the dates given.*

*The Antiquity of Man.* Sir Charles Lyell. 4th edition, 1873
*The Principles of Geology.* Sir Charles Lyell
*Prehistoric Times.* Sir John Lubbock. 5th edition, 1890
*The Origin of Civilisation.* Sir John Lubbock
*Early Man in Britain.* Prof. W. Boyd Dawkins
*Cave-hunting.* Prof. W. Boyd Dawkins
*Les Cavernes et leurs Habitants.* Par Julien Fraipont. (Paris, 1896)
*Geology.* Sir Joseph Prestwich. 2 vols.
*Collected Papers on Controversial Questions in Geology.* Sir Joseph Prestwich. 1895
*On Certain Phenomena . . . upon the Tradition of the Flood.* Sir Joseph Prestwich. 1895
*Prehistoric Europe.* Prof. James Geikie
*The Great Ice Age.* New edition, 1895. Prof. James Geikie
*Ice-work Past and Present.* 1896. Prof. T. G. Bonney
*Climate and Time.* Dr. James Croll
*Man and the Glacial Period.* G. F. Wright
*Ice Age in North America.* G. F. Wright. 1889
*Prehistoric Annals of Scotland.* 2 vols. Sir Daniel Wilson
*Prehistoric Man.* Sir Daniel Wilson
*The Lost Atlantis and other Ethnographical Studies.* 1892. Sir Daniel Wilson
*Man the Primeval Savage.* 1894. Worthington G. Smith
*Anthropological Studies.* 1891. Miss A. W. Buckland
*Reliquiæ Aquitanicæ.* H. Christy and E. Lartet. Edited by Prof. Rupert Jones
*Reliquiæ Diluvianæ.* Dean Buckland
*Flint Chips.* E. T. Stevens
*Jottings on Stonehenge.* E. T. Stevens (Brown, Salisbury)
*Ancient Stone Implements of Great Britain and Ireland.* Sir John Evans

*Ancient Bronze Implements of Great Britain and Ireland.* Sir John Evans
*Die Bronzezeit in Oberbayern.* Dr. Julius Nane (Munich). In progress. (This will be the most complete work on the subject.)
*Hommes Fossiles et Hommes Sauvages.* A. de Quatrefages
*Geology of England and Wales.* H. B. Woodward
*The Mammoth and the Flood.* 1887. Sir H. H. Howorth
*The Glacial Nightmare and the Flood.* 1893. Sir H. H. Howorth
*Man before Metals.* 1885. N. Joly
*The Story of Prehistoric Man.* 1895. Ed. Clodd
*Matériaux pour l'Histoire Positive de l'Homme.* Gabriel et A. de Mortillet. (This is now continued as a journal under the title *L'Anthropologie,* Paris.)
*The Deserts of Southern France.* 1894. Rev. S. Baring Gould. Vol. I. (Contains some new illustrations of engravings by the reindeer hunters) *Musée Préhistorique.* Gabriel et A. de Mortillet
*Der Mensch.* J. Ranke. 2 vols.
*Die Urgeschichte der Menschen.* M. Hoernes
*Der vorgeschichtliche Mensch.* Baer. Edited by Helvald
*Materialien zur Vorgeschicte der Menschen im östlichen Europa.* Kohn and Mehlis (Jena)
*L'Age du Renne* (in progress). Girod et Massénat (Paris)
*Exposition Universelle de Paris : Histoire du Travail et des Sciences Anthropologiques.* Section I (photos)
*Early History of Mankind.* Prof. E. B. Tylor
*Anthropology.* 1881. Prof. E. B. Tylor
*Primitive Culture.* Prof. E. B. Tylor
*L'Habitation Humaine.* 1895. C. Garnier (Paris)
*Prehistoric America.* 1885. Du Pouget (Marquis de Nadaillac)
*M... ners and Monuments of Prehistoric People.* 1892. Du Pouget (Marquis de Nadaillac)
*Human Origins.* 189 . Sam Laing. (Not very trustworthy)
*An Exploration of Dartmoor.* J. L. Warden Page
*A History of Egypt.* 1895. Prof. W. M. Flinders Petrie
*Stonehenge.* Prof. W. M. Flinders Petrie. (Most careful measurements.)
*The Sculptured Stones of Scotland.* John Stuart
*Rude Stone Monuments.* Sir James Ferguson. (Useful for illustrations)

*Stone Monuments.* J. B. Waring
*Industrial Arts of Denmark.* J. J. A. Worsaae
*Scandinavian Arts.* J. J. A. Worsaae
*La Création de l'Homme.* Edited by C. Flammarion. (Out of date, but has some artistic restorations)
*Les Origines.* J. H. Rosny, 'Collection Papyrus,' 1895. (Artistic, but not trustworthy)
*The Origin of Inventions.* 1889. A. A. Ellis (New York)
*Woman's Share in Primitive Culture.* 1894. O. T. Mason (New York)
*Le Costume Historique.* 1888. A. Racinet (Paris). (Fine coloured plates)
*Le Costume des Peuples Anciens et Modernes.* (Fine coloured plates). F. Hottenroth (Paris)
*Celtic Scotland.* W. F. Skene. 3 vols.
*The Dawn of Civilisation.* 1894. Maspero
*English Fairy Tales* (see Notes). 1890. J. Jacobs. (Illustrated by Batten)
*Scotland in Pagan Times (Bronze and Stone Ages).* Joseph Anderson
*Scotland in Early Christian and Pagan Times.* Joseph Anderson
*A Tour through the Islands of Orkney and Shetland.* Joseph Anderson
*Ancient Scottish Weapons.* Joseph Anderson
*Essai sur les Dolmens.* W. Bonstetten
*Primitive Inhabitants of Scandinavia.* Prof. Sven Nilsson. Edited by Sir J. Lubbock
*The Seven Sagas of Prehistoric Man.* James Stoddart
*British Barrows.* 1877. Canon Greenwell
*Handbook for Visitors to the Royal Irish Academy's Museum of Irish Antiquities.* Sir W. Wilde
*Ireland Past and Present* (a Lecture). Sir W. Wilde
*Finns, Fairies, and Picts.* 1893. David MacRitchie
*The Testimony of Tradition.* 1890. David MacRitchie
*Excavations on Cranbourne Chase.* 1887. Gen. Pitt-Rivers (Col. Lane Fox)
*The Lake Dwellings of Europe.* 1890. Dr. Robert Munro
*Rambles and Studies in Bosnia-Herzegovina and Dalmatia.* 1896. Dr. Robert Munro
*Prehistoric Man in Ayrshire.* 1896. John Smith
*Excavations at Carnac (Brittany).* James Miln
*The Lake Dwellings of Switzerland.* Dr. F. Keller

*Antiquités lacustres* (in progress). 1894. Lausanne: Lib. Bridel et Rouge. (This will be the most complete work. It is the work of two Swiss Societies, assisted by the Government of Vaud. One vol. just out)
*Primitive Folk.* Élisée Reclus

For general Ethnology the reader should consult Rätzel's *Völkerkunde* (2 vols. Leipzig), now appearing in English under the title 'The History of Mankind' (Macmillan and Co.), in 1*s.* monthly parts. The coloured plates are very well executed.

## JOURNALS &c.

*The Encyclopedia Britannica*
*Konversations-Lexicon.* Brockhaus
*Konversations-Lexicon.* Meyer
*British Association Reports*
*Quart. Journ. Geol. Soc.*
*Geological Magazine*
*Nature*
*Natural Science*
*Archæologia* (Soc. of Antiquaries, London)
*American Journal of Science*
*Journal of the Anthropological Institute*
*Proc. Society of Antiquaries,* Scotland
*Archiv für Anthropologie* (Brunswick)
*Zeitschrift für Ethnologie* (Berlin)
*Zeitschrift für Ethnologie* (Wien)
*Compte Rendu du Congrès International d'Anthrop. et d'Archéol. Préhist.,* 1873
*L'Anthropologie* (Paris)
*Revue Archéologique* (Paris)
*Jour. Wilts Arch. and Nat. Hist. Soc.* (Devizes)
*Trans. Royal Dublin Soc.*, vol. vi. 1896 (Dubois on ' Pilhecanthropus ')
*Trans. Royal Irish Academy*
*Jour. Ethnological Society*
*Archæological Review*

## BY THE SAME AUTHOR.

Cheap Edition, Revised and Enlarged. Large Crown 8vo. 6s.

# EXTINCT MONSTERS.

A POPULAR ACCOUNT OF SOME OF THE LARGER FORMS OF ANCIENT ANIMAL LIFE.

With Twenty-four Plates by J. SMIT, and numerous other Illustrations; and a Preface by Dr. HENRY WOODWARD, F.R.S.

### OPINIONS OF THE PRESS.

'If the popularity of extinct monsters has of late not been quite what it once was, it ought to be revived by Mr. Hutchinson's excellent book.'
— GUARDIAN.

'This is undoubtedly the best book that Mr. Hutchinson has written.'
— ATHENÆUM.

'A wonderfully interesting volume.'—REVIEW OF REVIEWS.

'Mr. Hutchinson is a skilful and trustworthy guide through the "valley of dry bones."'—DAILY CHRONICLE.

'His book, in short, is both attractive and useful, and will add to his reputation as a popular, but accurate, writer on geological subjects.'
— SATURDAY REVIEW.

'A piece of natural history that is far more amusing than most novels, and as full of instruction as a book of its size can well be.'
— NATIONAL OBSERVER.

'It is thoroughly readable.'—THE FIELD.

'This work is an admirable account of some of the larger forms of ancient animal life.'—NEW YORK NATION.

'The work fully sustains his reputation as a popular writer on scientific subjects.'—NEW YORK TIMES.

---

Cheap Edition. Large Crown 8vo. 6s.

# CREATURES OF OTHER DAYS.

With Twenty-four Plates by J. SMIT, and numerous other Illustrations; and a Preface by Sir W. H. FLOWER, Director of the Natural History Museum.

'Mr. Hutchinson, who is already well and favourably known as the author of "Extinct Monsters," has, in his new book, rendered a signal service to the cause of what is loosely called popular science. . . . Mr. Hutchinson combines scientific knowledge with the gift of popular exposition.'—DAILY NEWS.

'Mr. Hutchinson knows how to invest palæontology with interest for the ordinary reader.'—SPECTATOR.

'The style is clear and picturesque.'—MANCHESTER GUARDIAN.

'A work of profound learning and also of irresistible fascination.'
— THE WORLD.

---

LONDON: CHAPMAN & HALL, LIMITED.

# RURAL ENGLAND.

'*A series of books of really incomparable freshness and interest.*'—ATHENÆUM.
'*Books unsurpassed in power of observation and sympathy with natural objects by anything that has appeared since the days of Gilbert White.*'—DAILY NEWS.

## WORKS BY THE LATE RICHARD JEFFERIES.

**THE GAMEKEEPER AT HOME;** or, Sketches of Natural History and Rural Life. New Edition, with all the Illustrations of the former Edition. Crown 8vo. 5s.

'Delightful sketches. The lover of the country can hardly fail to be fascinated whenever he may happen to open the pages. It is a book to read and keep for reference, and should be on the shelves of every country gentleman's library.'—SATURDAY REVIEW.

**ROUND ABOUT A GREAT ESTATE.** New Edition. Crown 8vo. 5s.

'To read a book of his is really like taking a trip into some remote part of the country, where the surroundings of life remain very much what they were thirty or forty years ago. Mr. Jefferies has made up a very pleasant volume.'—THE GLOBE.

**WILD LIFE IN A SOUTHERN COUNTY.** New Edition. Crown 8vo. 6s.

'A volume which is worthy of a place beside White's "Selborne." In closeness of observation, in power of giving a picture far beyond the power of a mere word-painter, he is the equal of the Selborne rector—perhaps his superior. This is a book to read and to treasure.'
THE ATHENÆUM.

**THE AMATEUR POACHER.** New Edition. Crown 8vo. 5s.

'Unsurpassed in power of observation and sympathy with natural objects by anything that has appeared since the days of Gilbert White.'—DAILY NEWS.
'We have rarely met with a book in which so much that is entertaining is combined with matter of real practical worth.'—THE GRAPHIC.

**HODGE AND HIS MASTERS.** New Edition. Cr. 8vo. 7s. 6d.

'The one great charm of Mr. Jefferies' writings may be summed up in the single word "graphic." He has a rare power of description, and in "Hodge and his Masters" we find plenty of good reading.'—STANDARD.
'Mr. Jefferies knows his ground well and thoroughly, and writes with much of his wonted straightforwardness and assurance.... Pleasant and easy reading throughout.'—ATHENÆUM.

**WOODLAND, MOOR, AND STREAM;** being the Notes of a Naturalist. Edited by J. A. OWEN. Third Edition. Crown 8vo. 5s.

'As a specimen of word-painting, the description of the quaint old fishing village close to the edge of the North Kent marshes can hardly be surpassed.... The book is capitally written, full of good stories, and thoroughly commendable.'—THE ATHENÆUM.

**FOREST TITHES;** and other Studies from Nature. By the Author of 'Woodland, Moor, and Stream,' &c. Edited by J. A. OWEN. Crown 8vo. 5s.

'The book should be read. It is full of the spirit of the South Country, and as we read it we seem to hear again the clack of the millwheel, the cry of the water-fowl, and the splash of fish.'—SPECTATOR.

**ALL THE YEAR WITH NATURE.** By P. ANDERSON GRAHAM. Crown 8vo. 5s.

'Of the 28 papers composing the volume there is not one which does not brim over with love of Nature, observation of her by-paths, and power of sympathetic expression.'—OBSERVER.

London: SMITH, ELDER, & CO., 15 Waterloo Place.

www.ingramcontent.com/pod-product-compliance
Lightning Source LLC
Chambersburg PA
CBHW031856220426
43663CB00006B/647